Maria Gini, Richard Voyles (Eds.)

Distributed Autonomous Robotic Systems 7

Maria Gini, Richard Voyles (Eds.)

Distributed Autonomous Robotic Systems 7

With 120 Figures

Springer

Maria Gini,
Computer Science & Engineering
University of Minnesota 200 Union St SE
room 4-192 Mineapolis, MN 55455, USA

Richard Voyles,
Computer Science & Engineering
University of Minnesota 200 Union St SE
room 4-192 Mineapolis, MN 55455, USA

Library of Congress Control Number: 2006928857

ISBN-10 4-431-35878-1 Springer-Verlag Tokyo Berlin Heidelberg New York
ISBN-13 978-4-431-35878-7 Springer-Verlag Tokyo Berlin Heidelberg New York

Printed on acid-free paper

Springer is a part of Springer Science+Business Media
springer.com

Printing and binding: Hicom, Japan

Preface

The goal of the 8th Symposium on Distributed Autonomous Robotic Systems (DARS) is to exchange and stimulate research ideas to realize advanced distributed robotic systems. Technologies, algorithms, and system architectures will be presented and discussed during the symposium.

DARS 2006 builds upon past successes and provides an exciting environment for researchers to present and discuss their novel theoretical results, implementations, and applications. DARS successfully took place in 1992, 1994, and 1996 in Japan (Riken, Wako), in 1998 in Germany (Karlsruhe), in 2000 in Knoxville (Tennessee, USA), in 2002 at Fukuoka (Japan), and in 2004 at LAAS in Toulouse (France).

DARS 2006 will be held in the Minneapolis campus of the University of Minnesota, in the Electrical Engineering and Computer Science building.

A total of 42 technical papers were submitted by authors from multiple countries. All the submissions were rigorously reviewed by the Program Committee. Of those submissions 24 were accepted. The overall outcome of the revision process is an excellent selection of papers that showcase the research in distributed autonomous robotics today.

We would like to take this opportunity to thank everyone involved with the organization of DARS 2006. First, we would like to thank the members of the Program Committee, who did a thorough and conscientious job in reviewing a large number of papers. The members of the Advisory Commtitee provided invaluable help and support throughout the process of organizing the conference.

We warmly welcome all representatives from industry, government, and academia joining us in Minneapolis in July 2006.

Minneapolis, MN, USA
July 2006

Maria Gini
Richard Voyles

DARS 2006 Advisory Committee

Tamio Arai, University of Tokyo, Japan
Hajime Asama, University of Tokyo, Japan
Raja Chatila, LAAS, Toulouse, France
Rüediger Dillmann, University of Karlsruhe, Germany
Toshio Fukuda, Nagoya University, Japan
Lynne Parker, University of Tennessee, Knoxville, USA

DARS 2006 Program Committee

Julie Adams, Vanderbilt University
Mazda Ahmadi, University of Texas
Marcelo Ang, Jr., National University of Singapore
Yoshikazu Arai, Iwate Prefectural University
Minoru Asada, Osaka University
Silvia Botelho, FURG, Brazil
Zack Butler, Rochester Institute of Technology
Andres Castano, JPL, USA
Stephen Damer, University of Minnesota
Alexis Drogoul, IRD, France
Dominique Duhaut, Universite de Bretagne-Sud
Hugh Durrant-Whyte, University of Sydney
Patrick Fabiani, ONERA, France
Roderich Gross, Université Libre de Bruxelles
Akio Ishiguro, Nagoya University
Kuniaki Kawabata, RIKEN, Japan
Kazuhiro Kosuge, Tohoku University
Vijay Kumar, University of Pennsylvania
Daisuke Kurabayashi, Tokyo Institute of Technology
Monica LaPoint, University of Minnesota
Amy Larson, University of Minnesota
Alcherio Martinoli, EPFL, Switzerland
Bruce Maxwell, Swarthmore College
Satoshi Murata, Tokyo Institute of Technology
Anibal Ollero, University of Seville
Jun Ota, University of Tokyo
Enrico Pagello, University of Padova
Lucia Pallottino, Universita di Pisa
Ioannis Rekleitis, McGill University
Isabel Ribeiro, Instituto Superior Tecnico
Andy Russell, Monash University
Paul Rybski, Carnegie Mellon University
Alessandro Saffiotti, Orebro University

Behnam Salemi, USC/ISI
Frank Schneider, FGAN, Germany
Sanjiv Singh, Carnegie Mellon University
Sascha Stoeter, ETH, Switzerland
Ken Sugawara, Tohoku Gakuin University
Il Hong Suh, Hanyang University
Ichiro Suzuki, University of Wisconsin, Milwaukee
ZhiDong Wang, Tohoku University
Kjerstin Williams, Cal Tech
Heinz Wörn, Universitat Karlsruhe
Ning Xi, Michigan State University
Jizhong Xiao, City College of New York
Mark Yim, University of Pennsylvania
Eiichi Yoshida, AIST, Japan

Contents

List of Contributors

Julie A. Adams
Department of Electrical Engineering
& Computer Science
Vanderbilt University
Nashville, TN 37212, USA
julie.a.adamsg@vanderbilt.edu

Mazda Ahmadi
Department of Computer Sciences
The University of Texas at Austin
Austin, TX 78712, USA
mazda@cs.utexas.edu

Francesco Amigoni
Dipartimento di Elettronica e
Informazione
Politecnico di Milano
Piazza L. da Vinci 32
20100 Milano, Italy
amigoni@elet.polimi.it

Ronald C. Arkin
College of Computing
Georgia Institute of Technology
Atlanta, GA 30332, USA
arkin@cc.gatech.edu

Tucker Balch
College of Computing
Georgia Institute of Technology
Atlanta, GA, 30332, USA
tucker.balch@cc.gatech.edu

Dídac Busquets
Robotics Institute
Carnegie Mellon University
Pittsburgh, PA 15213, USA
busquets@eia.udg.es

Chee Kong Cheng
Cooperative Systems and Machine
Intelligence Lab
DSO National Laboratories
20 Science Park Drive
Singapore 118230
ccheekon@dso.org.sg

Nikolaus Correll
Swarm-Intelligent Systems Group
Ecole Polytechnique Fédérale
Lausanne, Switzerland
nicholaus.correll@epfl.ch

Joseph Djugash
Robotics Institute
Carnegie Mellon University
Pittsburgh, PA 15213, USA
robojoe@cmu.edu

Gregory Dudek
School of Computer Science
McGill University
Montreal, Quebec, Canada H3A 2A7
dudek@cim.mcgill.ca

Dominique Duhaut
VALORIA
University of South Britanny
56017 Vannes Cedex, France
Dominique.Duhaut@univ-ubs.fr

Yehuda Elmaliach
Computer Science Department
Bar Ilan University
Ramat Gan, Israel
elmaley@cs.biu.ac.il

Yoichiro Endo
College of Computing
Georgia Institute of Technology
Atlanta, GA 30332, USA
endo@cc.gatech.edu

Giulio Fontana
Dipartimento di Elettronica e Informazione
Politecnico di Milano
Piazza L. da Vinci 32
20100 Milano, Italy
fontana@elet.polimi.it

Toshio Fukuda
Dept. of Micro-Nano Systems Engineering
Nagoya University
Furo-cho, Chikusa-ku
Nagoya 464-8603, Japan
fukuda@mein.nagoya-u.ac.jp

Tetsuro Funato
Tokyo Institute of Technology
Ookayama 2-12-1, Meguro-ku
Tokyo 152-8552, Japan
funato@irs.ctrl.titech.ac.jp

Fabio Garigiola
Dipartimento di Elettronica e Informazione
Politecnico di Milano
Piazza L. da Vinci 32
20100 Milano, Italy
garigiol@airlab.elet.polimi.it

Maria Gini
Dept of Computer Science & Engineering
University of Minnesota
Minneapolis. MN 55455, USA
gini@cs.umn.du

Mark Gossage
Cooperative Systems and Machine Intelligence Lab
DSO National Laboratories
20 Science Park Drive
Singapore 118230
gmark@dso.org.sg

Claude Guéganno
VALORIA
University of South Britanny
56017 Vannes Cedex, France
claude.gueganno@univ-ubs.fr

Geoffrey Hollinger
Robotics Institute
Carnegie Mellon University
Pittsburgh, PA 15213, USA
gholling@andrew.cmu.edu

Wisnu Jatmiko
Dept. of Micro-Nano Systems Engineering
Nagoya University
Furo-cho, Chikusa-ku
Nagoya 464-8603, Japan
wisnu@robo.mein.nagoya-u.ac.jp

Boyoon Jung
NavCom Technology, Inc.
20780 Madrona Avenue
Torrance, CA 90503, USA
bjung@navcomtech.com

Nidhi Kalra
Robotics Institute
Carnegie Mellon University
Pittsburgh, PA 15213, USA
nkalra@cmu.edu

Akiya Kamimura
Intelligent Systems Institute
National Institute of Advanced Industrial Science and Technology (AIST)
Tsukuba, Japan
kamimura.a@aist.go.jp

Gal A. Kaminka
Computer Science Department
Bar Ilan University
Ramat Gan, Israel
galk@cs.biu.ac.il

Shigeru Kokaji
National Institute of Advanced Industrial Science and Technology (AIST)
Tsukuba, Japan
s.kokaji@aist.go.jp

Sarit Kraus
Computer Science Department
Bar Ilan University
Ramat Gan, Israel
sarit@cs.biu.ac.il

Daisuke Kurabayashi
Tokyo Institute of Technology
Ookayama 2-12-1, Meguro-ku
Tokyo 152-8552, Japan
dkura@irs.ctrl.titech.ac.jp

Haruhisa Kurokawa
Intelligent Systems Institute
National Institute of Advanced Industrial Science and Technology (AIST)
Tsukuba, Japan
kurokawa-h@aist.go.jp

Elizabeth Liao
Robotics Institute
Carnegie Mellon University
Pittsburgh, PA 15213, USA
eliao@andrew.cmu.edu

Luke Ludwig
Dept of Computer Science & Engineering
University of Minnesota
and BAESystems
Minneapolis, MN 55455, USA
ludwig@cs.umn.du

Alcherio Martinoli
Swarm-Intelligent Systems Group
Ecole Polytechnique Fédérale
Lausanne, Switzerland
alcherio.martinoli@epfl.ch

Maja J Matarić
Computer Science Department
University of Southern California
Los Angeles, CA 90089, USA
mataric@usc.edu

Colin McMillen
School of Computer Science
Carnegie Mellon University
Pittsburgh, PA 15213, USA
mcmillen@cs.cmu.edu

David Meger
School of Computer Science
McGill University
Montreal, Quebec, Canada H3A 2A7
dmeger@cim.mcgill.ca

Tsuyoshi Mizuguchi
Osaka Prefecture University
1-1, Gakuen-cho
Sakai, 599-8531, Japan
gutchi@ms.osakafu-u.ac.jp

Satoshi Murata
Graduate School of Science and Engineering
Tokyo Institute of Technology
Ookayama 2-12-1, Meguro-ku
Tokyo 152-8552, Japan
murata@dis.titech.ac.jp

Masahito Nara
Tokyo Institute of Technology
Ookayama 2-12-1, Meguro-ku
Tokyo 152-8552, Japan
m-nara@irs.ctrl.titech.ac.jp

Ai Peng New
Cooperative Systems and Machine Intelligence Lab
DSO National Laboratories
20 Science Park Drive
Singapore 118230
naipeng@dso.org.sg

Ioannis Rekleitis
Canadian Space Agency
Longueuil, Canada
Ioannis.Rekleitis@space.gc.ca

Avi Rosenfeld
Computer Science Department
Bar Ilan University
Ramat Gan, Israel
rosenfa@cs.biu.ac.il

Maayan Roth
Robotics Institute
Carnegie Mellon University
Pittsburgh, PA 15213, USA
mroth@andrew.cmu.edu

Sanem Sariel
Department of Computer Engineering
Istanbul Technical University
Istanbul, 34496, Turkey
sariel@cs.itu.edu.tr

Kosuke Sekiyama
Department of Human and Artificial Intelligence Systems
Fukui University 3-9-1 Bunkyo Fukui 910-850, Japan
sekiyama@dis.his.fukui-u.ac.jp

Dylan A. Shell
Computer Science Department
University of Southern California
Los Angeles, CA 90089, USA
shell@usc.edu

Reid Simmons
Robotics Institute
Carnegie Mellon University
Pittsburgh, PA 15213, USA
reids@cs.cmu.edu

Sanjiv Singh
Robotics Institute
Carnegie Mellon University
Pittsburgh, PA 15213, USA
ssingh@ri.cmu.edu

Jason Stack
Naval Surface Warfare Center
Panama City, FL, 32407 USA
Jason.stack@navy.mil

Peter Stone
Department of Computer Sciences
The University of Texas at Austin
Austin, TX 78712, USA
pstone@cs.utexas.edu

Ken Sugawara
Tohoku Gakuin University
2-1-1, Tenjinzawa, Izumi
Sendai, 981-3193, Japan
sugawara@cs.tohoku-gakuin.ac.jp

Gaurav S. Sukhatme
Computer Science Department
University of Southern California
Los Angeles, CA 90089, USA
gaurav@robotics.usc.edu

Yuzuru Terada
Graduate School of Science and Engineering
Tokyo Institute of Technology
Ookayama 2-12-1, Meguro-ku
Tokyo 152-8552, Japan
string@mrt.dis.titech.ac.jp

Kohji Tomita
National Institute of Advanced Industrial Science and Technology (AIST)
Tsukuba, Japan
tomita@aist.go.jp

Patrick Ulam
College of Computing
Georgia Institute of Technology
Atlanta, GA 30332, USA
pulam@cc.gatech.edu

Manuela Veloso
School of Computer Science
Carnegie Mellon University
Pittsburgh, PA 15213, USA
veloso@cs.cmu.edu

Lovekesh Vig
Department of Electrical Engineering & Computer Science
Vanderbilt University
Nashville, TN 37212, USA
flovekesh.vig@vanderbilt.edu

Alan R. Wagner
College of Computing
Georgia Institute of Technology
Atlanta, GA 30332, USA
alan.wagner@cc.gatech.edu

A Distributed Biconnectivity Check

Mazda Ahmadi and Peter Stone

Department of Computer Sciences,
The University of Texas at Austin,
{mazda,pstone}@cs.utexas.edu
http://www.cs.utexas.edu/~{mazda,pstone}

Summary. For many distributed autonomous robotic systems, it is important to maintain communication connectivity among the robots. That is, each robot must be able to communicate with each other robot, perhaps through a series of other robots. Ideally, this property should be robust to the removal of any single robot from the system. In this work, we define a property of a team's communication graph that ensures this property, called biconnectivity. We present a distributed algorithm to check if a team of robots is biconnected, prove its correctness, and analyze it theoretically.

1.1 Introduction

Many applications of distributed autonomous robotic systems can benefit from, or even may require, the team of robots staying within communication connectivity. For example, consider the problem of multirobot surveillance [1, 2], in which a team of robots must collaboratively patrol a given area. If any two robots can directly communicate at all times, the robots can coordinate for efficient behavior. This condition holds trivially in environments that are smaller than the robots' communication range. However in larger environments, the robots must actively maintain physical locations such that any two robots can communicate — possibly through a series of other robots. Otherwise, the robots may lose track of each others' activities and become miscoordinated. Furthermore, since robots are relatively unreliable and/or may need to change tasks (for example if a robot is suddenly called by a human user to perform some other task), in a stable multirobot surveillance system, if one of the robots leaves or crashes, the rest should still be able to communicate. Some examples of other tasks that could benefit from any pair of robots being able to communicate with each other, are space and underwater exploration, search and rescue, and cleaning robots.

We say that robot R_1 is *connected* to robot R_2 if there is a series of robots, each within communication range of the previous, which can pass a message from R_1 to R_2. In order for the team to stay connected, it must be the case that every robot is connected to each other robot either directly or via *two* distinct paths that don't share any robots in common. We call this property *biconnectivity*: the removal of any one robot from the system does not disconnect the remaining robots from each other.

In previous work, we developed algorithms for multirobot surveillance under the assumption that each pair of robots could communicate directly [2]. This communication assumption enabled the robots to negotiate to achieve an efficient task division, but it constrained us to small environments. This work is the first attempt to extend these algorithms to larger environments.

To the best of our knowledge, the problem of enabling robots to remain connected in the face of robot failures has not been explored before. Typical related work in graph theory is on algorithms to find a biconnected component in a graph with optimal time complexity (e.g. [3]), in dynamic graphs (e.g. [4]), or in a restricted subclass of all graphs (e.g. [5]). In all these cases, the algorithms are either centralized, or if distributed, each node has full knowledge of the whole graph. Some work in distributed computing is closer in spirit to our work, however a main difference between their problem statement and ours is that in distributed computing (e.g. [6, 7]), any node can send a message to any other node. That is, the nodes are not restricted to send messages only through existing edges of the graph.

We tackle this problem by dividing it into three main steps: (1) Checking whether a team of robots is *currently* biconnected, (2) Maintaining biconnectivity should a robot be removed from (or added to) the team, and (3) Constructing a biconnected multi-robot structure from scratch. To be applicable for teams of autonomous robots, all algorithms must be fully distributed.

In this paper we focus on fully achieving and analyzing Step 1. Steps 2 and 3 remain as future work. Note that it is possible to achieve steps 2 and 3, even if inelegantly, by having the robots move back to a base and disperse from there whenever they find that they are no longer biconnected.

For the purposes of this paper, we assume that robots have constant and identical communication ranges. This assumption applies in the case of homogeneous robot teams (or at least teams with homogeneous transmitters) such that the range is not dependent on a robot's battery level. This assumption allows us to assume the connection graph among robots is undirected: if robot A can send a message to robot B, then the reverse is also true. Extension of this work to the case where robots have heterogeneous communication capabilities is also a part of our future work plans.

After the introduction, in Section 1.2 graph theory background and assumptions about the investigated multirobot systems is presented. Section 1.3 presents distributed algorithms to detect if the robots are biconnected. Finally Section 1.4 concludes the paper.

1.2 Preliminaries

We first provide some graph definitions and theorems which will be used later in the paper. For basic graph definitions, such as vertex, edge, neighbor, path and loop please see [8]. Later in the section, definitions and assumptions which are specific to our multirobot system will be presented.

Definition 1. Internally vertex-disjoint paths. *Two paths between v_1 are v_2 are internally vertex-disjoint if they have no vertices in common except v_1 and v_2.*

Definition 2. Biconnected graph. *If in graph G, after removing any vertex, it is possible to find a path from any vertex to any other one, the graph G is said to be biconnected.*

Definition 3. Doubly connected vertices. *In graph G, we say vertex v_1 and v_2 are doubly-connected iff there are two or more internally vertex-disjoint paths between v_1 and v_2.*

Lemma 1. *Undirected graph $G(V, E)$ is biconnected if and only if any two vertices $v_1, v_2 \in V$ are doubly-connected.*

Proof. It is a special case of Menger's Theorem (See Theorem 3.3.1 of [8]).

Note that in undirected graphs one vertex being doubly-connected to all other vertices is not a sufficient condition for the graph to be biconnected. For an example see Figure 1.1 where v is doubly-connected to all other vertices, but removing v makes the graph disconnected. In the following theorem we show that in an undirected graph if there are two vertices that are doubly-connected to all other vertices, then the graph is biconnected.

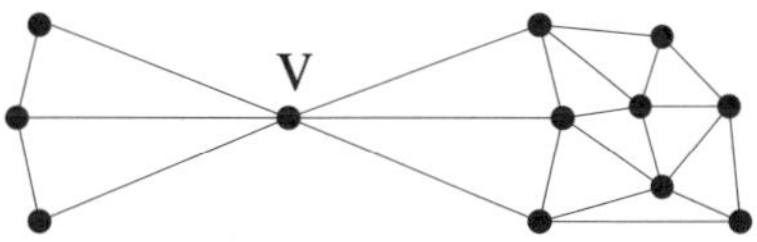

Fig. 1.1. V is doubly-connected to all other vertices, but the graph is not biconnected.

Theorem 1. *Undirected graph $G(V, E)$ is biconnected if and only if there exists two distinct vertices $v_1, v_2 \in V$ such that both v_1 and v_2 are doubly-connected to any other vertex in V.*

Proof. If Graph $G(V, E)$ is biconnected, then by Lemma 1 any two vertices are doubly-connected to all other vertices.

It remains to prove that if there exists $v_1, v_2 \in V$ ($v1 \neq v2$) such that they are doubly-connected to all other vertices in V, then $G(V, E)$ is biconnected. Assume $v_i \in V$ is removed from V. We must show that the graph remains connected. If $v_i = v_1$, then since any other vertices $v_j \in V$ were doubly-connected to v_2, the graph remains connected. Similarly for $v_i = v_2$. Now assume $v_i \neq v_1, v_2$. Since every other vertex $v_j \in V$ was doubly-connected to v_1, and v_i is in at most one of the two paths between v_j and v_1, after removal of v_i, v_j remains connected to v_1. Thus the graph remains connected after removal of any vertex: it is biconnected.

We look at our multirobot system as a graph, such that its vertices are robots and edge $(v_1 v_2)$ exists in the graph iff the robot corresponding to v_1 can communicate directly to the robot corresponding to v_2 (i.e. v_1 and v_2 are in communication range of each other). A formal definition of robot graph follows.

Definition 4. Robot graph $RG(V, E)$ *is a graph, where its vertices (V) are the robots and $(v_1, v_2) \in E$ iff corresponding robots to v_1 is a neighbor of corresponding robot to v_2. Size of V (i.e. number of robots in the multirobot team) is called n in this paper.*

Assumption 1 *Robots are aware of the maximum number of robots in the system, which can be considerably higher than the actual number of robots. The maximum number is called N throughout the paper.*

Assumption 2 *Robots have identical communication capabilities.*

As a result of the above assumption, the neighbor property is symmetric, and the robot graph is undirected.

Definition 5. Connected. *We say robot R_1 and robot R_2 are connected, when in the corresponding robot graph, there is a path between R_1 and R_2.*

Assumption 3 *Each robot has a unique and ordered ID. For robot X its ID is called X.id.*

Next, the definition and assumption regarding the communication between robots is provided.

Definition 6. Message, stamped message. *For our purposes, a* message, *which is used for robot communication, is a string in format* $(T,(S))$, *where* T *indicates the type of the message, and* S *is a list of robot* stamps. *Message* $(T,(S))$ *is said to be* stamped *by robot R iff* $R.id \in (S)$. *Robot* R *stamps message* $(T,(S))$ *by generating new message* $(T,(S,R.id))$.

Assumption 4 *When called for by the protocol, robots relay messages to one another. Robots start processing received messages, as soon as they get them. The maximum period from the time that robot* R_1 *receives message* X, *until its neighbor robot* R_2 *receives the processed (possibly stamped) version of message* X *from* R_1 *is c seconds.*

1.3 Algorithms to Check Biconnectivity

As mentioned in Section 1.1, checking for biconnectivity is the first step towards the overall goal of achieving and maintaining a biconnected multirobot structure. It is an important step, because the robots must be able to detect if they are not biconnected, so that they can take remedial actions to restore *biconnectivity* before they lose *connectivity*. The remedial actions could be as simple as all robots moving back to a base and dispersing from there.

Note that the biconnected property is a global property of the multirobot system: robots cannot determine whether or not it holds from purely local information. For example see Figure 1.1, where the graph is not biconnected, and the robots associated with the nodes on the right side of the graph need global information about the nodes on the left side to know that the whole structure is not biconnected.

In our approach, each robot R, maintains two lists:

- CR_R *(connected robots)*: the list of robots that are connected to R.
- DCR_R *(doubly-connected robots)*: the robots doubly-connected to R.

Each robot R first fills the CR_R list, then using that, the DCR_R list is computed. Finally with the help of the DCR_R list, it detects if the robot graph is biconnected.

In the rest of this section, we first provide an algorithm (CR-FILL) for filling CR_R in Section 1.3.1, then another algorithm (DCR-FILL) is presented in Section 1.3.2 which fills the DCR_R lists with the help of the already computed CR_R lists. Afterwards in Section 1.3.3, an algorithm which checks the biconnectivity with the help of the computed DCR_R lists is provided. All these algorithms are distributed and each robot runs them independently. Finally an analysis of the presented algorithms is provided in Section 1.3.4.

1.3.1 CR-FILL

In this subsection, we provide an algorithm for filling the CR_R list. That is for robot R, it finds the robots that are connected to it.

The basic idea is for the robots to stamp and pass messages in the system. In this way, if there is a path of $r_0 \rightarrow r_1 \rightarrow r_2 \rightarrow R$, between r_0 and R, robot R will receive a message that is stamped by r_0, r_1 and r_2. Thus it will know that it is connected to those robots, and will add them to the CR_R list.

Two helper algorithms must run continuously on all the robots to help the CR-FILL algorithm. The first helper algorithm dictates how messages should be passed around. In the second, if robot r has not sent a message for a while and another robot is running a CR-FILL algorithm and needs a stamped message initiated from r, it will send a stamped message. After introducing these two helper algorithms, the CR-FILL algorithm itself will be presented.

Using the first helper algorithm, all robots continually stamp and pass biconnected type messages that they receive. Any robot r, which receives $(\text{“}CR\text{”}, (S))$, checks the content of S, and if $r.id \notin S$ it stamps the message and send it out. That is, it sends message $(\text{“}CR\text{”}, (S, r.id))$. If $r.id \in S$, it does not send any message because stamping and sending it would lead to a duplicate ID in $(S, R.id)$. For an overview of this algorithm see Algorithm 1.

Algorithm 1 Message passing algorithm which robots continually run

1: **upon** receiving a message of form $(\text{“}CR\text{”}, (S))$ **do**
2: **if** $R.id \notin (S)$ **then** robot R broadcast message $(\text{“}CR\text{”}, (S, R.id))$.
3: **end upon**

Any robot, upon receiving a message of format $(\text{“}CR\text{”}, S)$, if it has not sent out a $(\text{“}CR\text{”}, (R.id))$ message in the last Nc seconds (Recall from Assumption 1, that N is the maximum number of robots in the system), it sends out message $(\text{“}CR\text{”}, (R.id))$ (see Algorithm 2).

Algorithm 2 The condition for initiating a “CR” message.

1: **upon** receiving a message of form $(\text{“}CR\text{”}, (S))$ **do**
2: **if** has not sent out a $(\text{“}CR\text{”}, (R.id))$ message in the last Nc seconds **then**
3: broadcast message $(\text{“}CR\text{”}, (R.id))$
4: **end if**
5: **end upon**

When calling the main CR-FILL algorithm, robot R starts by initializing the CR_R list to empty. Afterwards each time it receives message $(\text{“}CR\text{”}, (S))$, it adds all the IDs in S to CR_R. While still adding IDs to the CR_R, at time Nc, robot R sends out a stamped message $(\text{“}CR\text{”}, (R.id))$. The pseudocode of this algorithm is available in Algorithm 3.

Algorithm 3 Pseudocode for the CR-FILL algorithm for robot R

1: Time 0 (start of the algorithm): initialize the CR_R to empty.
2: Time Nc: broadcast message $(\text{“}CR\text{”}, (R.id))$
3: **if** message of form $(\text{“}CR\text{”}, (S))$ is received **then** add IDs in (S) to CR_R.

Since the length of the longest path in the graph is less than N, the maximum time for a message to reach robot r_2 from r_1 is Nc seconds. We now show any robot r that is connected to R will be added to CR_R within $3Nc$ seconds. Any robot r that is connected to R, receives the stamped message from robot R within $2Nc$ seconds (note that the first message is sent at time Nc). If it has sent a message in the last Nc seconds, robot R has gotten that message. Otherwise, it will send out a message which will be heard by R in at most Nc seconds. Thus after $3Nc$ seconds, CR_R represents the correct list of robots that are connected to R. This analysis is based on the assumption that the robots do not change connectivity in the $3Nc$ seconds that CR-FILL runs. Note that after $3Nc$ seconds, no message is left in the system. Because message $(\text{“}CR\text{”}, (S))$ can only survive if it is received by robots such that their ID is not in (S), $2Nc$ seconds after sending the first message any remaining message in the system has been stamped by all robots, and it cannot survive any longer. Also note that c is ideally on the order of milliseconds, though in practice it may be difficult to guarantee such small bounded transmission times. In such cases, the algorithms as is may become impractical for large teams of fast-moving (so that connectivity changes quickly) robots.

1.3.2 DCR-FILL

In this subsection, DCR-FILL, an algorithm to fill the DCR lists is presented. It is assumed the message passing algorithm (Algorithm 1) is running continually by all robots.

The basic idea for filling DCR_R for robot R is to find the robots that are in a common loop with R. When the robot graph is undirected (Assumption 2), there is a loop including both R and R' iff two internally vertex-disjoint paths (Definition 1) exist between R and R'. In this case, R and R' are doubly-connected (Definition 3). According to Algorithm 1, robots pass stamped messages around. When robot R receives a message that has been stamped by itself (i.e. R), it knows the robots that have stamped the message after the R stamps are in a common loop with R, and should be added to DCR_R.

Robot (r) starts by broadcasting message $(\text{“}DCR\text{”}, (r.id))$, which will be heard by all of its neighbor robots. Upon receiving message $(\text{“}DCR\text{”}, (S))$, if this is the first time to receive a “DCR” message, it resets DCR_r to empty (initializing), afterwards it checks the content of (S). If its own ID *is* in the stamp part of the message (S), it can represent (S) as $(S_1, r.id, S_2)$. If S_2 includes more than one vertex, it means that there is a loop and the robot adds all the IDs in S_2 to DCR_r. If S_2 includes only one vertex, it means that the robot has got back a message from a robot that it has just sent a message to, and should be ignored. Algorithm 4 presents the pseudocode of this algorithm.

We now show that for robot R, the DCR-FILL algorithm sets the correct DCR_R list within nc seconds.

Theorem 2. *For any robot R, the* DCR-FILL *algorithm finds the full list of doubly-connected robots (DCR_R) within nc seconds.*

Proof. Consider the robot graph $RG(V, E)$ for the robots, and also $v_1 \in V$ represents robot R. To prove this theorem, we need to show for any vertex $v_2 \in V$, if v_1 is doubly-connected to v_2, then $v_2 \in DCR_{v_1}$, and if ($v_2 \in DCR_{v_1}$) then v_2 and v_1 are doubly-connected. Also we need to show DCR-FILL *is completed (i.e. there is no message in the system) after nc seconds.*

We start with the first part, and assume v_1 and v_2 are doubly-connected, so there are two internally vertex-disjoint paths (a loop) between them. The starting message

Algorithm 4 Pseudocode for DCR-FILL algorithm

Time 0: robot R broadcasts $(\text{"}DCR\text{"}, (R.id))$.
upon receiving a message of form $(\text{"}DCR\text{"}, (S))$ **do**
  if this is the first time to receive a "DCR" message **then**
    reset DCR_R to empty
  end if
  if $R.id \in (S)$ **then**
    split (S) to $(S_1, R.id, S_2)$
    if $size(S_2) > 1$ **then** add the IDs in S_2 to DCR_R
  end if
end upon

from v_1 will go through the loop, and vertex v_1 will get back the message that it stamped earlier, which is now also stamped by v_2. Thus, v_2 will be added to DCR_{v_1}.

Now we have to prove the other part, assuming $v_2 \in DCR_{v_1}$. Based on the algorithm, the only way that v_2 is added to DCR_{v_1} is when v_1 receives a message $(\text{"}DCR\text{"}, (S_i, v_1.id, S_j, v_2.id, S_k))$. Notice that based on the condition in the algorithm ($size(S_2) > 1$ (Algorithm 4), if both S_j and S_k are empty, the IDs will not be added to DCR_R, also recall that there is no duplicate IDs in messages, because no robot stamps a message that it has previously stamped. The two internally vertex-disjoint paths between v_1 and v_2 are $v_1 S_j v_2$ and $v_1 S_k v_2$. Thus v_1 and v_2 are doubly-connected.

Similar to the argument at the end of Section 1.3.1, after nc seconds no message remains in the system, and the algorithm terminates.

1.3.3 Biconnectivity Check

After running CR-FILL and DCR-FILL consecutively, the CR and DCR lists will be accurate. Notice that both algorithms for filling the CR and DCR lists finish within a known time limit. Thus the robots should wait $3Nc$ seconds, and afterwards CR and DCR lists will be accurate. For robot r if CR_r and DCR_r are equal, it means that r is doubly-connected to all the robots that it is connected to. By Theorem 1, we know if there are two robots r_1 and r_2 that are doubly-connected to all other robots, then the robot graph is biconnected. Also, we know by Lemma 1 that if there is a robot that is not doubly-connected to all other robots, the robot graph is not biconnected. Thus if the robot and one of its neighbors is doubly-connected to all other robots, the robot knows that the robot graph is biconnected. Also if the robot or one of its neighbors is not doubly-connected to all other robots, it will know that the robot graph is not biconnected.

The overview of the biconnectivity check algorithm is shown in Algorithm 5. The *initiator* robot (which can be any robot who wants to check biconnectivity) starts by sending a ("CHECK-REQUEST",()) message to its neighbors to ask them to check if they are doubly-connected to other robots or not. Upon receiving a ("CHECK-REQUEST",()) the other robots run biconnectivity check (unless they are already running it) as non-initiators (skipping line 3 of Algorithm 5). Note that multiple robots can run the biconnectivity check algorithm in parallel.

If the robot is doubly-connected to all other robots, it sends the message ("DC-TRUE", ()) to all its neighbor robots, and a ("DC-FALSE",()) message otherwise. If the robot is doubly-connected to all other robots and receives a ("DC-TRUE", ()) message, it knows that the robot graph is biconnected. Otherwise (if it is not biconnected to all other robots, or receives a ("DC-FALSE", ()) message) it knows that the robot graph is not biconnected. Since the initiator and its neighbors should run the biconnectivity check, the total time needed for the biconnectivity check to complete is $6Nc + 2c$ seconds.

Algorithm 5 Pseudocode for biconnectivity check algorithm. It returns *true* if the robot graph is biconnected, and *false* otherwise.

run CR-FILL and DCR-FILL in parallel, and wait $3Nc$ seconds for them to be finished.
if (initiator) **then** send message ("CHECK-REQUEST",())
if $size(DCR_R) = size(CR_R)$ **then**
    send message ("DC-TRUE",())
else
    send message ("DC-FALSE",())
    return false;
end if
if a message of form ("DC-FALSE", ()) is received **then** return false;
if a message of form ("DC-TRUE", ()) is received **then**
    if $size(DCR_R) = size(CR_R)$ **then** return true;
end if

1.3.4 Algorithms' Analysis

In this section we analyze both the time and communication complexity of the CR-FILL and DCR-FILL algorithms.

CR-FILL, DCR-FILL and biconnectivity check algorithms use $3Nc$, nc, and $6Nc+2c$ seconds to complete respectively.

For the analysis of the number of messages, we assume that the time needed for the message sent by a robot to reach its neighbor is constant (c). This assumption does not change the total number of messages, but possibly can change the maximum number of messages at any point in time. The worst case for the number of messages in the multirobot system happens when the robot graph is fully connected, that is every two robots are neighbors. For both DCR-FILL and CR-FILL algorithms, the maximum number of messages in the system is $n!$, because when robot R receives message $(T, (S))$, where S has size of i, the message only survives if robot R sends it to the robots that have not already stamped the message, and $n - i - 1$ such robots exist. Thus from the messages that have started from robot R, $(n - 1)!$ can exist in the system, and since each of the n robots starts one message of its own, at any point in time, the maximum number of messages in the system for DCR-FILL or CR-FILL algorithm is $n!$.

Note that the times provided above are the worst case, and especially when there are many robots, the robot graph is most likely not fully connected. An example of a still densely connected robot graph with 35 robots is given in Figure 1.2, there in each time period, on average each robot deals with 1037 messages, which is a manageable number. But 1037 messages is still a lot to process in each time step.

Our analysis is based on the assumption that CR-FILL must generate special-purpose messages. When messages are being sent for other purposes, the stamps required can simply be appended to those, thus eliminating the need for *many* extra messages. Though if each message does not normally generate a broadcast responses, *some* extra messages may be needed in order to keep the time complexity the same. In principle, DCR-FILL only needs to be started by 2 robots, thus reducing the number of messages required by a factor of $\frac{2}{n}$. When those two robots have computed their DCR_R, they can let the others know if the robot graph is biconnected. The technical details of how to determine which two robots send the starting messages is beyond the scope of this paper. However in essence, it is similar to maintaing team leaders, which is a common practice in multirobot systems (e.g. [9]).

If CR-FILL uses existing messages and DCR-FILL is run by only 2 robots, the total number of messages is $2(n-1)!$ in the worse case, and for the graph of Figure 1.2, on average each robot deals with approximately 69 messages, which is an easily manageable number in most realistic scenarios.

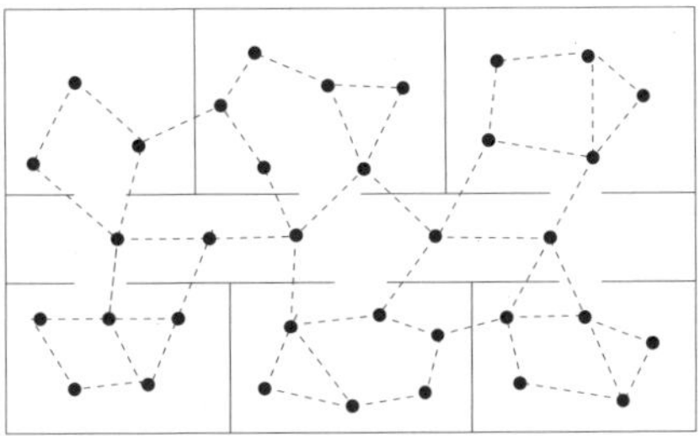

Fig. 1.2. An example of a common robot graph with 35 robots.

All the presented algorithms only store CR and DCR lists, which have a maximum size of n. Thus all algorithms use $O(n)$ memory space.

The time complexity discussed here is for each received message. That is, we assume the decisions are made when messages arrive. Both the CR-FILL and DCR-FILL algorithms have time complexity of $O(n)$ because they only traverse a list of IDs, which has size of at most n. The time complexity of biconnectivity check, which include both CR-FILL and DCR-FILL is of $O(2n)$.

1.4 Conclusion and Future Work

In this paper, we defined and argued the need for biconnected multirobot structures. A distributed algorithm for checking biconnectivity is presented, proven correct, and analyzed theoretically.

In future work, we aim to provide optimality bounds for the provided checking biconnected algorithms. The assumption that robots have identical communication capabilities should be relaxed, which will result in the robot graph being directed. Also, our algorithms for maintaining biconnectivity and constructing biconnected structure from scratch remains as future work.

Acknowledgments

We would like to thank Kurt Dresner and Nick Jong for their valuable comments on an earlier version of this paper. This research was supported in part by NSF CAREER award IIS-0237699 and ONR YIP award N00014-04-1-0545.

References

1. Parker, L.E.: Distributed algorithms for multi-robot observation of multiple moving targets. Autonomous Robots **12** (2002) 231–255

2. Ahmadi, M., Stone, P.: A multi-robot system for continuous area sweeping tasks. In: Proceedings of International Conference on Robotics and Automation (ICRA), to appear. (2006)
3. Tarjan, R., Vishkin, U.: Finding biconnected componemts and computing tree functions in logarithmic parallel time. In: 25th Annual Symposium on Foundations of Computer Science, 1984. (1984) 12–20
4. Westbrook, J., Tarjan, R.E.: Maintaining bridge-connected and biconnected components on-line. Algorithmica (Historical Archive) **7** (1992) 433–464
5. Galil, Z., Italiano, G.F.: Maintaining biconnected components of dynamic planar graphs. In: Proceedings of the 18th International Colloquium on Automata, Languages and Programming, London, UK, Springer-Verlag (1991) 339–350
6. Swaminathan, B., Goldman, K.J.: An incremental distributed algorithm for computing biconnected components (extended abstract). In: Proceedings of the 8th International Workshop on Distributed Algorithms, London, UK (1994)
7. Ahuja, M., Zhu, Y.: An efficient distributed algorithm for finding articulation points, bridges, and biconnected components in asynchronous networks. In: Proceedings of the Ninth Conference on Foundations of Software Technology and Theoretical Computer Science, London, UK, Springer-Verlag (1989) 99–108
8. Diestel, R.: Graph Theory. Springer, New York (1997)
9. *R. Alur et al.*: A framework and architecture for multirobot coordination. In: Seventh International Symposium on Experimental Robotics. (2001)

A Method for Building Small-Size Segment-Based Maps

Francesco Amigoni, Giulio Fontana, and Fabio Garigiola

Dipartimento di Elettronica e Informazione, Politecnico di Milano,
Piazza Leonardo da Vinci 32, 20133 Milano MI, Italy
{amigoni,fontana}@elet.polimi.it, garigiol@airlab.elet.polimi.it

Summary. Segment-based maps have recently emerged as an effective solution to reduce the dimensions of environment models built by mobile robots. In this paper, we present a novel method for building segment-based maps that contain a small number of line segments. The method works also when data are collected by many robots. Experimental results show that our approach is effective in significantly reducing the size of the resulting maps.

1 Introduction

Robotic mapping addresses the problem of acquiring spatial models of physical environments through mobile robots [12]. Multirobot mapping has attracted attention because of both the robustness and the efficiency of exploring in parallel with multiple robots. Maps can be represented topologically (e.g., by graph-based data structures) or geometrically (e.g., by data structures storing grids, points, or line segments). Segment-based maps have been recently advocated as a way to reduce the dimensions of the data structures storing the representation of the environment [2, 3, 7, 13]. If, on the one hand, information can be extracted more efficiently from segment-based maps, on the other hand, the advantages are effective only when the number of line segments is kept as small as possible to avoid redundancy in data.

In this paper, we propose a novel method for building segment-based maps composed of a small number of line segments. Our method incrementally integrates a newly acquired scan S with an existing map M_t. It aligns S with M_t on the basis of the estimated pose (i.e., position and rotation) from which S has been acquired; then, it fuses the line segments of S and M_t to reduce the complexity of the resulting map M_{t+1}. Our method for fusing line segments constitutes the main original contribution of this paper. Experimental results show that our approach is effective in significantly reducing the number of line segments in the final map.

Our method does not make any assumption about how the scans S are collected. This makes our approach naturally applicable to cases in which the scans are acquired by multiple mobile robots, possibly equipped with different sensors. Actually, in a multirobot scenario, the need of reducing the number of line segments in the final map is even more pressing, since the robots, exploring independently, may acquire redundant information about the same portions of the environment. With multiple mobile robots, the scans can be sent to a *map manager* that, upon receiving a scan S, integrates it with the previous map. Although a centralized map manager can constitute a bottleneck for the scalability of the system, it appears to be the simplest solution when few mobile robots are involved and communication is reliable.

This paper is organized as follows. In the next section, our map building approach is illustrated. Section 3 presents the proposed method for fusing line segments. Section 4 shows some experiments that validate our method. Section 5 compares our method with other methods for segment-based maps. Section 6 concludes the paper.

2 Our Map Building Approach

We provide a method for map building that integrates a newly acquired scan S and an existing map M_t into an updated map M_{t+1}, trying to keep the number of line segments in M_{t+1} as small as possible. A *scan* is the result of a sensing operation and it is assumed to be a collection of line segments in a 2D space. The map M_t, being built by integrating a number of scans, is composed of line segments. M_t is updated whenever a new scan is available, thus t is not intended to represent absolute time but the number of sensing operations (and of consequent integrations). We do not make any assumption about how and by which robot the scans are acquired. This makes our approach naturally applicable to scenarios in which multiple robots equipped with different sensors explore an environment. In our experiments, we used laser range finders but, in principle, cameras and other sensors can be used as long as they can provide segment-based scans. We assume that all the relevant obstacles are at the height of sensor perception and, in the case of multiple sensors, we assume that all the sensors perceive the environment at the same height.

Our map building method works in three steps: scan acquisition, scan alignment, and scan fusion. In the following of this section, we briefly describe the first two steps. Scan fusion is presented in the next section. We first introduce a preliminary definition. The *distance* between two line segments s and s' is calculated as $D(s, s') = \min(d(s, s'), d(s', s))$, where $d(s, s') = \max_{p \in s}(\min_{p' \in s'}(||p - p'||))$, and $||\cdot||$ is the Euclidean distance. Note that $D(s, s')$ is similar to, and always less than, the Haussdorff distance ($H(s, s') = \max(d(s, s'), d(s', s))$, often used in computer graphics and in computer vision.

Scan acquisition produces a scan S. In our experimental activity, we used a SICK LMS 200 laser range scanner operating at a height of approximatively 50 cm. For each scan, the sensor acquires an ordered (counterclockwise) sequence of 361 distance measurements along directions separated by 0.5°, sweeping 180°. The raw data returned by the sensor can thus be seen as a set of (ordered) points expressed in polar coordinates, with the origin of the coordinate frame in the sensor itself. These points are processed to obtain line segments: they are first clustered and then the points in each cluster are approximated by a line segment following a technique similar to that used in [5]. The main differences are: firstly, our clusters contain points that are approximated by a single line segment instead of a polyline; secondly, we fit the approximating line segment using a least mean square algorithm and calculating the distance between a point and a line segment along the line connecting the sensor to the point. In this way, we better cope with the errors of laser range scanners.

Each scan S is associated with the estimated pose P_S (in the reference frame of the map M_t) from which it has been acquired. In our case, the robots maintain knowledge about their pose using data coming from odometry. However, it is well-known that odometry is unreliable. In order to refine the estimated poses, after each scan acquisition, we apply a scan alignment algorithm derived from that presented in [9]. That algorithm aligns two scans composed of points and iteratively performs two steps. Firstly, it finds the pairs of corresponding points belonging to the two scans. Secondly, it calculates the transformation (i.e., rotation and translation) that minimizes the distances between the corresponding points. We adapted this algorithm to align the scan S and the map M_t, both composed of line segments. The main issue of our algorithm is how to establish that two line segments correspond to each other. Two line segments *correspond* when their angle is smaller than T_θ and their distance is smaller than T_D (in our experiments, we used the following thresholds: $T_\theta = 5°$ and $T_D = 10\,\text{cm}$). Our *scan alignment* algorithm aligns S and M_t in the following way.

1. Represent S in the reference frame of M_t, by applying to S the transformation given by the estimated pose P_S.
2. Find the rotation $\bar{\theta}$ that, when applied to S, minimizes the distance between S and M_t:

$$\mathcal{D}(S, M_t) = \frac{1}{N_m + N_n} \cdot \left(\sum_{i=1}^{N_m} D'(s_i) + N_n \cdot T_D\right) \tag{1}$$

   where N_m is the number of line segments of S that correspond to at least a line segment in M_t and N_n is the number of line segments of S that do not correspond to any line segment in M_t. In general, given a line segment s of S there is a set C_s of line segments of M_t that correspond to s (according to the definition of correspondence given above). Note that C_s can be also empty. $D'(s_i)$ is the average distance of line segment s_i (of S) to the line segments in C_{s_i}, weighted according to their lengths.
3. Apply the rotation $\bar{\theta}$ to S.
4. Find the translation $(\bar{x}, \bar{y})$ that, when applied to S, minimizes $\mathcal{D}(S, M_t)$.
5. Apply the translation $(\bar{x}, \bar{y})$ to S.
6. Repeat from step 2 until a given number of iterations or until $\mathcal{D}(S, M_t)$ is below a threshold.

We assume that the estimated pose P_S is precise enough to make two line segments representing the same portion of the environment correspond when calculating $\mathcal{D}(S, M_t)$ with (1). "Precise enough" is difficult to quantify exactly since it depends on the geometry of the environment. In our experiments, estimated (initial) poses with errors up to 10 cm and 15° were "precise enough". At the end of the above algorithm, S and M_t are aligned and their line segments can be fused.

3 The Proposed Method for Scan Fusion

In this section we present the proposed method for scan fusion that represents the main original contribution of this paper.

We represent a line segment s with the following tuple (see also Fig. 1):

$$s = \langle \theta, \rho, (x_1, y_1), (x_2, y_2), \sigma_\theta, \sigma_\rho \rangle \tag{2}$$

where θ and ρ are the parameters (angle and distance from origin, respectively) of the line supporting s, (x_1, y_1) and (x_2, y_2) are the coordinates of the extreme points of s, and σ_θ and σ_ρ represent the uncertainty of θ and ρ, respectively. When, as in our experimental activity (see the scan acquisition step in the previous section), a line segment s is obtained from points acquired by a laser range scanner, σ_θ and σ_ρ are calculated as follows:

$$\sigma_\theta = \arcsin\left(\frac{\varepsilon}{L/2}\right) \qquad \sigma_\rho = \varepsilon$$

where ε is the maximum distance between s and the points that "generated" it and L is the length of s (in our experiments $\varepsilon = 20\,\text{mm}$). This representation for line segments is based on that of [4]. We use the parameters σ_θ and σ_ρ instead of the parameter j of the original formulation, that represented the number of points from which the line segment has been derived. By eliminating j, our method is applicable also to cases in which line segments are not obtained from approximation of points (e.g., when they are obtained processing images captured by cameras), provided that σ_θ and σ_ρ are calculated properly.

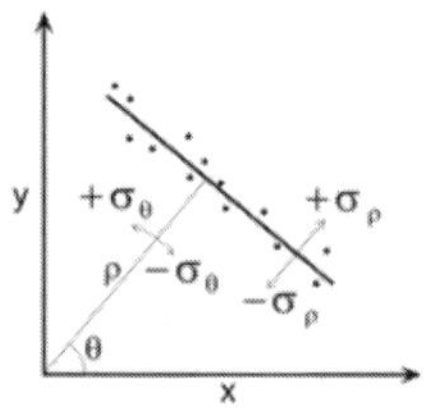

Fig. 1. The parameters of a line segment

We now illustrate the basic method for fusing two line segments s and s'; later we will extend it to a set of line segments. To decide whether s and s' can be fused, three conditions, called *fusion conditions*, must be satisfied:

1. the angle between s and s' is smaller than T_θ: $\widehat{ss'} < T_\theta$,
2. the distance between s and s' is smaller than T_D: $D(s, s') < T_D$,
3. the projections P_s and $P_{s'}$ of s and s' on their bisector (i.e., on the line that bisects the angle formed by s and s') overlap: $P_s \cap P_{s'} \neq \emptyset$.

Note that the first two conditions above define the correspondence of line segments (as introduced in the previous section).

When two line segments $s = \langle \theta, \rho, (x_1, y_1), (x_2, y_2), \sigma_\theta, \sigma_\rho \rangle$ and $s' = \langle \theta', \rho', (x'_1, y'_1), (x'_2, y'_2), \sigma'_\theta, \sigma'_\rho \rangle$ can be fused, namely when they satisfy all the above conditions, we calculate the resulting line segment $s^f = \langle \theta^f, \rho^f, (x^f_1, y^f_1), (x^f_2, y^f_2), \sigma^f_\theta, \sigma^f_\rho \rangle$ in the following way. The parameters of the line supporting s^f are:

$$\theta^f = \frac{\omega_\theta \cdot \theta + \omega'_\theta \cdot \theta'}{\omega_\theta + \omega'_\theta} \qquad \rho^f = \frac{\omega_\rho \cdot \rho + \omega'_\rho \cdot \rho'}{\omega_\rho + \omega'_\rho} \tag{3}$$

where the weights ω_θ and ω_ρ are defined as:

$$\omega_\theta = \frac{L}{L+L'} \cdot \frac{1}{\sigma_\theta^2} \qquad \omega_\rho = \frac{L}{L+L'} \cdot \frac{1}{\sigma_\rho^2}$$

where L and L' are the length of s and s', respectively. In a similar way, ω'_θ and ω'_ρ can be defined. The weights of a line segment are directly proportional to its length and inversely proportional to its uncertainty. This reflects the idea that long line segments are more "reliable" than short line segments.

The extreme points (x_1^f, y_1^f) and (x_2^f, y_2^f) of s^f are calculated by projecting the extreme points of s and s' on the line with parameters θ^f and ρ^f and by selecting, among the four projected points, the pair of farthest points.

Finally, the uncertainty σ_x^f (either σ_θ^f or σ_ρ^f) of s^f is calculated as shown in Fig. 2, that represents the behavior of σ_x^f as a function of L and L'. When $L \gg L'$, σ_x^f is approximatively equal to σ_x (s "absorbs" s'); conversely, when $L \ll L'$, σ_x^f is approximatively equal to σ'_x (s is "absorbed" by s'). When $L = L'$, σ_x^f is calculated as:

$$\sigma_x^f = \sqrt{\left[\frac{1}{\sigma_x^2} + \frac{1}{\sigma_x'^2}\right]^{-1}}$$

Namely, the uncertainty of the final segment s^f is reduced with respect to the uncertainty of both s and s'. In our experimental activity, we used a linearization of the curve of Fig. 2.

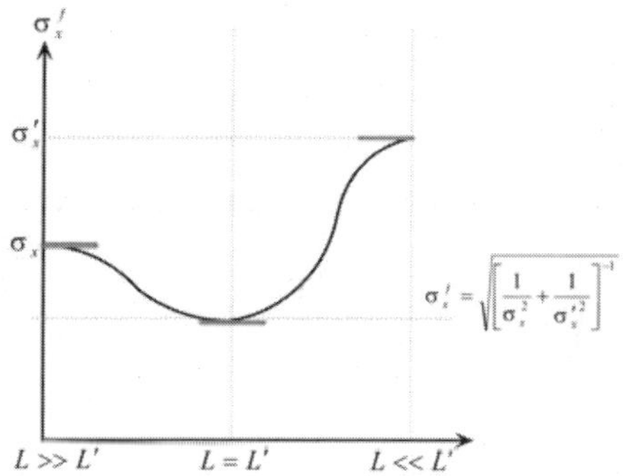

Fig. 2. The function σ_x^f

In order to reduce as much as possible the number of the line segments in the resulting map M_{t+1}, we apply the above fusion procedure to a set of line segments. More precisely, given M_t and S, acquired and aligned as explained in the previous section, we build, for each line segment s in S, the set C_s that contains all the line segments in M_t that can be fused with s (i.e., that satisfy the three fusion conditions stated above). Then, we approximate the line segments in $C_s \cup \{s\}$ with a line segment s^f, that is inserted in M_{t+1} (together with the line segments in M_t that do not correspond to any line segment in S). It is straightforward to extend the formulas (3) for θ^f and ρ^f to the case of many line segments. Similarly, also the extreme points (x_1^f, y_1^f) and (x_2^f, y_2^f) can be easily calculated in the case of many line segments. To calculate σ_θ^f and σ_ρ^f we use the following (approximate) iterative procedure. We order the line segments in $C_s \cup \{s\}$ from the longest one to the shortest one: $s_0, s_2, \ldots, s_n$. Then, we calculate $\sigma_x^{0,1} = \Phi(\sigma_x^0, \sigma_x^1)$, where Φ is the function of Fig. 2 (recall that x stands for either θ or ρ). At this point, we calculate $\sigma_x^{0,1,2} = \Phi(\sigma_x^{0,1}, \sigma_x^2)$ and so on until $\sigma_x^f = \sigma_x^{0,1,\ldots,n} = \Phi(\sigma_x^{0,1,\ldots,n-1}, \sigma_x^n)$ is obtained. In this way, the uncertainty of the longer line segments in $C_s \cup \{s\}$ has a stronger impact on the values of σ_θ^f and σ_ρ^f.

4 Experimental Results

In order to validate the proposed map building approach, we performed several experiments with a mobile robot (based on a Robuter platform) equipped with a SICK LMS200 laser range scanner. The experiments have been conducted in different wall-bounded environments (both indoor and outdoor). We manually moved the robot through a sequence of observation points, with an average distance between observation points of about 800 mm. For each observation point we collected two scans. For each scan S, we applied our method to fuse S with the current map M_t to obtain M_{t+1}. We compared the maps obtained with our approach with maps obtained following our approach *without* performing the scan fusion of Section 3. Due to space limitations, in this section we report only some of the results we obtained.

In the first experiment we drove the robot along a corridor closed at one end. Along the path, 26 scans have been acquired. Fig. 3 shows the final maps obtained without scan fusion (composed of 151 line segments) and with scan fusion (composed of 21 segments). The reduction in the number of line segments is 86%. Fig. 4 shows the number of line segments in the two maps as the scans are acquired.

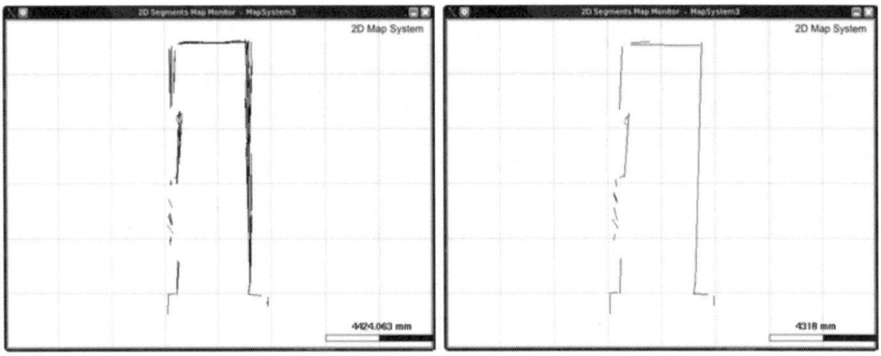

Fig. 3. First experiment: maps obtained without (left) and with (right) scan fusion

To preliminarily test our method in a multirobot scenario, we simulated the presence of two mobile robots. Given an environment, we initially drove our robot along a path, storing the acquired scans and odometry data in a log file. Then, we used the same robot to follow another path in the same environment. This time, as

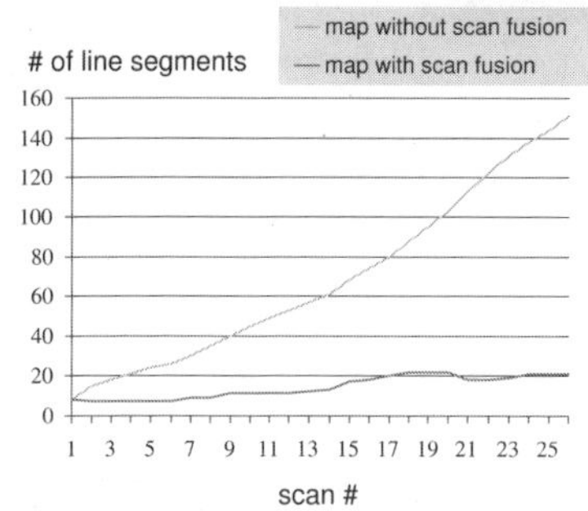

Fig. 4. Number of line segments vs. number of scans

the robot was acquiring scans along the second path, we applied our map building method to integrate these scans interleaved with those read from the log file. Overall 20 scans have been acquired: 8 along the first path and 12 along the second path. Fig. 5 shows the maps built (without scan fusion) integrating the scans collected along the two paths. Fig. 6 (left) shows the map obtained with scan fusion. The number of line segments in this map is reduced of 89% with respect to total number of line segments in the 20 scans (and in the maps of Fig. 5). Fig. 6 (right) shows the number of line segments as the scans are acquired.

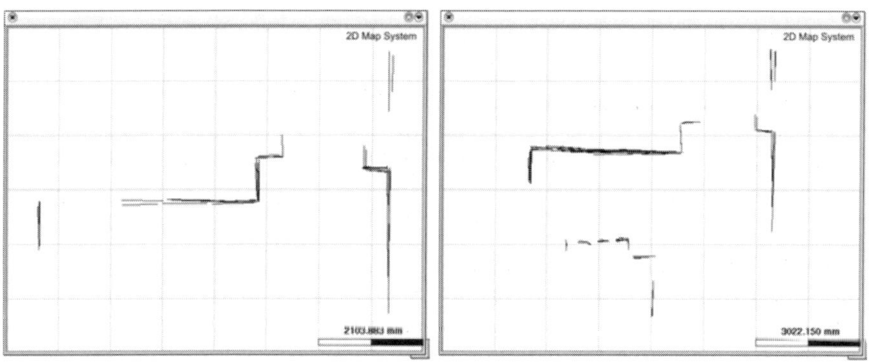

Fig. 5. Second experiment: maps obtained without scan fusion

We show another multirobot experiment. We simulated the second robot as illustrated above, collecting a total of 33 scans in an environment. Fig. 7 shows the final map obtained with scan fusion and the number of line segments as scans are acquired. In this case, the reduction in the number of line segments, with respect to the number of line segments in the 33 scans, is 85%.

The results presented above constitute a preliminary validation of the effectiveness of our approach. The proposed method dramatically reduces the number of line segments in the maps without loss of significant features of the environment. In our experiments, the reduction rates have ranged from 80% to 90%. Not surpris-

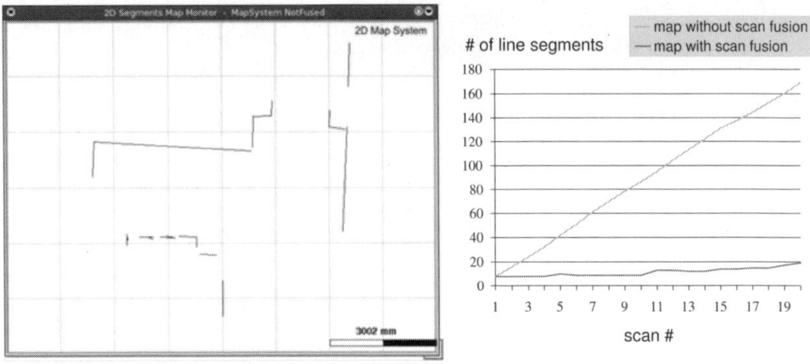

Fig. 6. Second experiment: map obtained with scan fusion (left) and number of line segments vs. number of scans (right)

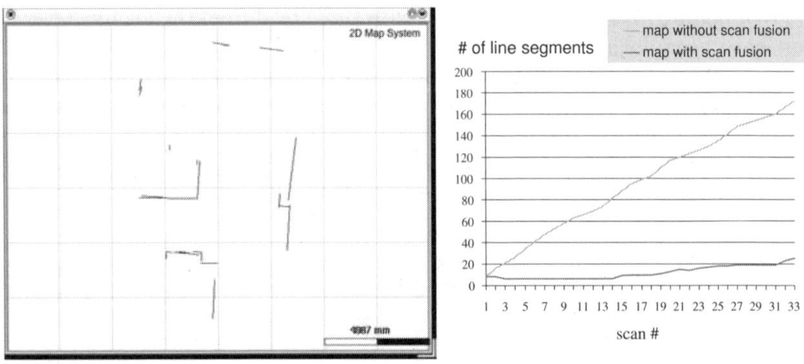

Fig. 7. Third experiment: map obtained with scan fusion (left) and number of line segments vs. number of scans (right)

ingly, the best results have been obtained in environments with long straight walls. In cluttered environments, a lot of small line segments approximate the irregular boundaries of objects; these line segments are difficult to fuse together in longer line segments, complicating the reduction of the dimensions of the maps (see the left wall of the corridor in Fig. 3, where a bicycle was present). The values of the parameters T_θ and T_D used as thresholds in the fusion conditions proved to be highly critical. We experimentally found satisfactory values; an analytical study could help in finding optimal values.

5 Related Works

In this section, we compare our approach to those presented in literature to build segment-based maps. Some approaches keep all the line segments in the final map without attempting any fusion, or applying limited fusion procedures (e.g., fusing collinear segments [5]).

In [10], line segments are represented by their extreme points. The authors propose a procedure to merge a newly acquired line segment s_n and an old line segment s_o. The procedure finds a line segment that approximates the measured points that form s_n and the points obtained by sampling s_o. The solution proposed in [14] describes a line segment using the center of gravity of its points and the direction θ of its supporting line. Line segments are grouped in clusters and, for each cluster, a new line segment is generated. Its parameters are the weighted average of the parameters of the line segments in the cluster. In this case, the weights are the variances of the center of gravity. Both these methods are applicable only when line segments are obtained starting from sensors that perceive points. The method in [13] calculates the probability that two line segments can be fused according to their angle, distance, and overlapping, in a way similar to our fusion conditions. However, it is not clear how the line segments are merged. In [7], the authors propose to evaluate whether two polylines can be fused on the basis of the similarity of their convex arcs. The fusion of two polylines is performed simulating a laser range scanner and so reducing the problem to the fusion of points. In [1], close line segments form matching chains, according to a measure of distance similar to that used in this paper. A matching chain is then approximated by a polyline with an iterative procedure.

There are some approaches using the so-called *fuzzy line segments* that provide a representation for line segments that intrinsically includes uncertainty. A simple fuzzy approach is proposed in [8], but it is suitable only for highly structured environments with perpendicular walls. A more complex solution appears in [4] where, as already discussed, line segments are defined similarly to (2). In this method, uncertainty values are used to decide if two or more line segments can be grouped in a cluster; whereas the numbers of measured points used to build line segments are used to merge the line segments of a cluster. A similar representation of line segments is used in [11], where a chi-squared criterion is used to identify line segments that can be merged. The resulting line segment is calculated as the average of the merged line segments, weighted by their accuracies. Clustering and fusion of fuzzy line segments are used also in [6] with a two-step process: local clustering (to group scan points and extract line segments by a fuzzy c-means algorithm) and global clustering and fusion (to reduce the total number of line segments of the map).

When compared with the approaches discussed above, the method proposed in this paper explicitly addresses the problem of reducing the number of line segments in the final map. In [2] a similar problem is considered, but with emphasis on the selection of the proper geometric primitives (line, arc, or cubic segments) to reduce the dimensions of maps. Moreover, our method is applicable to situations in which the line segments are not obtained starting from points. Finally, while the approaches for segment-based maps presented in literature are mainly oriented to single robot systems, our method is naturally applicable to multirobot systems.

6 Conclusions

In this paper we have presented a novel method that builds small-size segment-based maps. Experimental results validated the effectiveness of our approach in significantly reducing the number of line segments in the maps. The importance of the proposed method in a multirobot scenario lies in two main considerations. Firstly, it supports the building of small-size maps that can be easily managed and transmitted over low-bandwidth networks of mobile robots. Secondly, it is independent from the way in which the scans are acquired and, in principle, it is naturally scalable as the number of robots increases, since scans are integrated one at a time.

Future research will address the testing of our method in large environments and the use of different sensors mounted on different robots to collect the scans; we are currently working on scan acquisition performed by cameras. In this way, we aim to analyze the dependence of the parameters to the type of sensors and to the type of environments. Another future research direction is to consider techniques used in computer graphics for segment and polygon reduction.

References

1. F. Amigoni, S. Gasparini, and M. Gini. Scan matching without odometry information. In *Proc. of the Int'l Conf. on Informatics in Control, Automation and Robotics*, pages 349–352, 2004.
2. D. Austin and B. McCarragher. Geometric constraint identification and mapping for mobile robots. *ROBOT AUTON SYST*, 35:59–76, 2001.
3. E. Brunskill and N. Roy. SLAM using incremental probabilistic PCA and dimensionality reduction. In *Proc. of the IEEE Int'l Conf. on Robotics and Automation*, pages 342–347, 2005.
4. J. Gasós and A. Martín. Mobile robot localization using fuzzy maps. In T. Martin and A. Ralescu, editors, *Fuzzy Logic in AI - Proc. of the IJCAI'95 Workshop*, volume LNAI 1188, pages 207–224. Springer-Verlag, 1997.
5. H. H. Gonzáles-Baños and J. C. Latombe. Navigation strategies for exploring indoor environments. *INT J ROBOT RES*, 21(10-11):829–848, 2002.
6. Y. L. Ip, A. B. Rad, K. M. Chow, and Y. K. Wong. Segment-based map building using enhanced adaptive fuzzy clustering algorithm for mobile robot applications. *J INTELL ROBOT SYST*, 35:221–245, 2002.
7. L. J. Latecki, R. Lakaemper, X. Sun, and D. Wolter. Building polygonal maps from laser range data. In *Int'l Cognitive Robotics Workshop*, 2004.
8. M. López-Sánchez, F. Esteva, R. López De Màntanaras, and C. Sierra. Map generation by cooperativa low-cost robots in structured unknown environments. *AUTON ROBOT*, 5:53–61.
9. F. Lu and E. Milios. Robot pose estimation in unknown environments by matching 2D range scans. *J INTELL ROBOT SYST*, 18(3):249–275, 1998.
10. R. Mázl and L. Přeučil. Building a 2d environment map from laser range-finder data. In *Proc. of the IEEE Intelligent Vehicles Symposium*, pages 290–295, 2000.
11. J. W. Burdick S. T. Pfister, S. I. Roumeliotis. Weighted line fitting algorithms for mobile robot map building and efficient data representation. In *Proc. of the IEEE Int'l Conf. on Robotics and Automation*, pages 1304–1311, 2003.
12. S. Thrun. Robotic mapping: A survey. In G. Lakemeyer and B. Nebel, editors, *Exploring Artificial Intelligence in the New Millenium*. Morgan Kaufmann, 2003.
13. E. Zalama, G. Candela, J. Gómez, and S. Thrun. Concurrent mapping and localization for mobile robots with segmented local maps. In *Proc. of the IEEE/RSJ Int'l Conf. on Intelligent Robots and Systems*, pages 546–551, 2002.
14. L. Zhang and B. Ghosh. Line segment based map building and localization using 2D laser rangefinder. In *Proc. of the IEEE Int'l Conf. on Robotics and Automation*, pages 2538–2543, 2000.

Learning when to Auction and when to Bid

Dídac Busquets and Reid Simmons

Robotics Institute
Carnegie Mellon University
Pittsburgh, PA 15213, USA
{didac,reids}@cs.cmu.edu

Summary. The market based approach is widely used to solve the problem of multirobot coordination. In this approach, communication and computation costs are key issues, but have not been carefully addressed by the different architectures in the literature. In this paper, we present a method to reduce these costs, by adding the capability to learn whether a task is worth offering up for auction and also whether it is worth bidding for the task, based on previous experience about successful and unsuccessful bids. We show that the method significantly decreases communication and computation costs, while maintaining good overall performance of the team.

1 Introduction

In the past few years, the field of mobile robotics has begun to consider problems involving groups of robots. A team of robots can achieve tasks much more efficiently than using only one robot, and it has a wide range of applications, including planetary exploration, search and rescue, and surveillance. However, having multiple robots adds complexity to the problem, since some means of coordinating them is needed. To this end, many researchers in the field have focused their efforts on the topic of multirobot task allocation (MRTA). The problem posed by MRTA is: given a set of robots and a set of tasks to be executed, which tasks should be assigned to which robot in order to achieve the team's goal as effectively as possible?

This problem can be solved using different techniques, including those from operations research, scheduling, combinatorial optimization, and economic-based. In this paper we focus on the latter, the economic or market based approach. The main idea of this approach is to consider the tasks to be accomplished as goods, and the robots as self-interested agents that bid in auctions for these goods. As result of the auctions, the tasks are distributed among the robots and, if the robots bid rationally, the task distribution is the one that gets the most profit for the overall team. Moreover, once a robot is assigned a task, it can auction it off to the other robots, and it can also bid for tasks offered by other robots, if selling or getting those tasks is profitable.

Two key issues in this approach are the communication and computation costs for holding auctions and computing bids. Communication costs refer to the number of messages needed for running the auctions. Obviously, the fewer number of messages, the better, since the network resource in a team of robots is often limited. Computation costs refer to the computational cost of running the auctions, which consists of the bidder's cost of computing the bid for each task, and the auctioneer's cost for clearing the auctions.

In this paper, we present a method to reduce these costs, by adding the capability to learn which tasks to include in auctions and which tasks to submit bids for. The first reduces the number of tasks the bidders need to evaluate, the number of bids for the auctions, and the sizes of the call for bids. The second reduces the number of messages and the number of bids the auctioneer receives, since the bidders send fewer bids. Our empirical results demonstrate significant savings in both communication and computational costs, with little effect on the overall performance of the system.

2 Related work

Market-based mechanisms have been widely used in the multiagent community, and in the past few years the field of robotics has borrowed these ideas to solve multirobot problems. Examples of such architectures include those of Dias and Stentz [2], Gerkey and Matarić [4] and Golfarelli et al. [7]. However, as Gerkey and Matarić [5] point out, none of the existing implementations have carefully taken into account the communication and computational costs, which can have a major impact on the system's performance. Although they give a formal analysis of the cost of different implementations, they do not address how the costs could be reduced in order to improve the performance.

Gage and Murphy [3] addressed this problem using an emotion-based approach. They extended Parker's ALLIANCE architecture [10] adding a *shame emotion* to the robots, which controls the willingness of each robot to respond to a call for help from another one. This shame level increases as a function of the task being announced (namely, as a function of the time needed to arrive at the task location). However, the final decision of whether the robot is going to respond is based on a fixed threshold, no matter what task has caused the shame level to reach this threshold. This differs from our approach, in which the task characteristics directly drive the robot's decision of bidding or not.

We should point out that in our work the robots learn *when* to bid and not *what* to bid (i.e. the amount of the bid). The latter approach has been widely studied, and it focuses on learning bidding strategies that lead to maximizing the profit of an agent in an economic system [8, 9]. In contrast, our approach focuses on learning the probability of whether a given bid may win an auction. Similarly, our approach also learns *when to auction*, by learning the likelihood that a task being offered by a robot will be awarded to any of the other robots.

3 The market architecture

Before describing how we have addressed the problem of communication and computational costs, we give a brief overview of the market based architecture on which we have based our work [2].

Consider a team of robots assembled to perform a set of tasks, each of which has a specific reward for its completion, where each robot in the team is modeled as a self-interested agent and the team of robots is modeled as an economy. The goal of the team is to complete the tasks successfully while maximizing overall profit. Each robot aims to maximize its individual profit. However, since all revenue is derived from satisfying team objectives, the robot's self-interest is equivalent to maximizing global team objectives.

Internally, each robot is controlled by a three-layered architecture (see [2] for more details). In this paper we focus on the Planning layer, which is responsible for holding and participating in auctions to allocate tasks. It has two main components: a *Trader* that participates in the market, auctioning and bidding on tasks [2], and a *Scheduler* that determines task feasibility and costs for the Trader and interacts with other layers for task execution [1].

The tasks are introduced in the system by the OpTraders (Operator Traders), which receive high-level commands from human operators and translate them into task descriptions the robots can recognize. These tasks are then auctioned so that the robots bid for them. Upon receiving a Call for Bids, the Trader sends the task (or set of tasks) to the Scheduler, which computes the cost of executing it and the bid that should be sent back to the auctioneer. Basically, a bid is the amount a robot is willing to pay to get the opportunity to perform a task and later collect the corresponding reward. The Scheduler computes the bid as the difference between the profit obtained by including the task into the current schedule and the profit obtained if the robot does not execute this task. This calculation is computationally expensive, since the Scheduler needs to solve a hard optimization problem. Therefore, the fewer tasks it has to schedule (or reschedule), the faster it can compute the value of the bid. The auctioneer awards the task to the highest bidder, as long as that bid increases the auctioneer's overall profit (for the OpTraders this is always the case, since any bid will produce some profit). When the winning bidder completes the task, it informs the auctioneer and collects the reward.

4 Learning the probabilities

In most current implementations of the market-based approach, bidders respond to all the tasks being announced by the auctioneers. This massive response causes a large communication overhead that could be reduced if the bidders bid only for some of the tasks being offered. Obviously, if bidders are going to be selective about what bids to make, they should bid only for those tasks that they are most likely to get awarded.

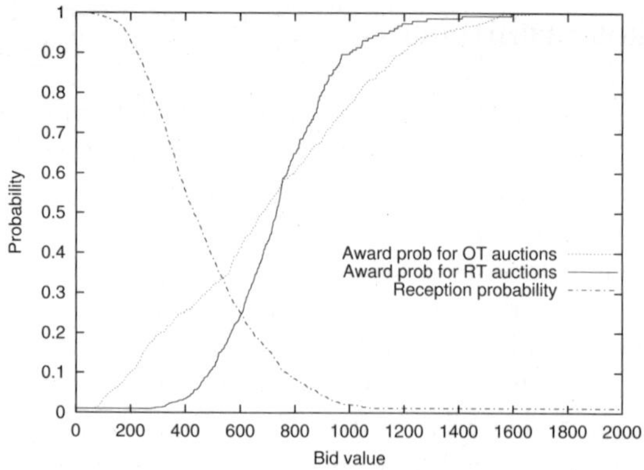

Fig. 1. Award and reception probabilities

Moreover, a robot can also try to auction the tasks it has been assigned, but has not yet executed, in order to increase its profit by trading them. In most market based models, the robot would try to auction all its pending tasks, which also incurs a large communication cost. Not only this, but it also forces the other robots to evaluate each of these tasks (increasing their computational cost) and send their bids (increasing again the communication load, and the computational cost for the auctioneer to clear the auction). These costs could be reduced if the robots offered only some of their remaining tasks. Again, one would prefer the robot to offer only those tasks that are most likely to be successfully bid for by some other robot (that is, receiving a bid that makes the trade profitable and, therefore, the auctioneer can award the tasks).

In the following sections, we present the heuristics that bidders and auctioneers use to determine whether to bid for a task and whether to offer a task to the other robots, respectively. These heuristics are based on the probability of a bid being the winner of an auction, and the probability of a bidder sending a successful bid for a task being offered. These probabilities are estimated by the robots from previous experience.

4.1 Award probability

In order to reduce the communication cost, the bidder should bid only for a few of the tasks being auctioned. When deciding whether to bid, the bidder should take into account the chances of the bid being awarded. To do so, the probability of a given bid being awarded in an auction is computed, which is then used by the bidder to decide whether to send a bid to the auctioneer. A uniform random number is generated, and if it is less than or equal to the probability of a bid of that value being awarded, the bid is sent. Thus, even bids with a low probability may eventually be bid for.

First, note that the auctions held by the OpTrader (OT) are different from those held by the robots (RT) because the OT does not have a minimum bid value for trading a task; it always makes some profit by accepting any bid. Thus, we learn two probability distributions, $AwProb^{OT}$ and $AwProb^{RT}$. With $AwProb^{OT}$ the bidders try to learn the probability of a bid being the winner of an auction (that is, the highest bid), while with $AwProb^{RT}$ the bidders try to learn the probability of a bid being higher than the minimum bid required by the auctioneer to trade a task.

These probabilities are computed as follows. Let WB^{OT} be the set of winning bids for OT auctions, WB^{RT} the set of winning bids for RT auctions, and MLB the set of the maximum losing bid for each RT auction (only when there is no winning bid, i.e. no bid was above the minimum price). Since the OT always awards a task if bid for, there is no such set for OT auctions. The probability of a bid of value b being awarded is computed as:

$$AwProb^{OT}(b) = \frac{won^{OT}(b)}{|WB^{OT}|} \qquad AwProb^{RT}(b) = \frac{won^{RT}(b)}{won^{RT}(b) + lost(b)}$$

where, $won^{a}(b) = |(x \in WB^{a} | x \leq b)|$ and $lost(b) = |(x \in MLB | x \geq b)|$.

That is, the award probability for OT auctions is the percentage of successful auctions whose winning bid is below than or equal to b, while for RT auctions it is the percentage of winning bids lower than or equal to b, with respect to all auctions in which a bid of b would definitely win or lose the auction. Note that there are RT auctions where a bid b is below the winning bid, but is not counted as a losing bid since it still may have been above the auctioneer's minimum bid price (which is private information).

Figure 1 shows the award probabilities for OT and RT auctions computed using these formulas (the bid values are in the interval 0-2000 because in our experiments the reward of each task is 2000, thus this is the maximum possible bid). As expected, the higher the bid, the higher the chances of being awarded. It can also be observed that for low bid values (below 500) the award probability for OT auctions is much higher than that for RT auctions, due to the fact that an OT does not impose a minimum value for trading a task.

4.2 Probability of auctioning a task

As discussed above, the number of tasks auctioned by the robots also increases the communication load of the system and the computational cost for the bidders. Thus, a robot should offer only a small percentage of its tasks. The decision of whether to auction a given task should be affected by the likelihood of the task being finally awarded to some other robot. A task auctioned by a robot (the auctioneer) is awarded to another one (the bidder) if the bid sent by the bidder is greater than the loss incurred by the auctioneer by giving away the task. In other words, an auctioneer trades a task only if it gets more profit by doing so than by keeping the task for itself.

Thus, the auctioneer should use the probability of receiving a bid greater than the loss of not executing a task to decide whether to offer it to the rest

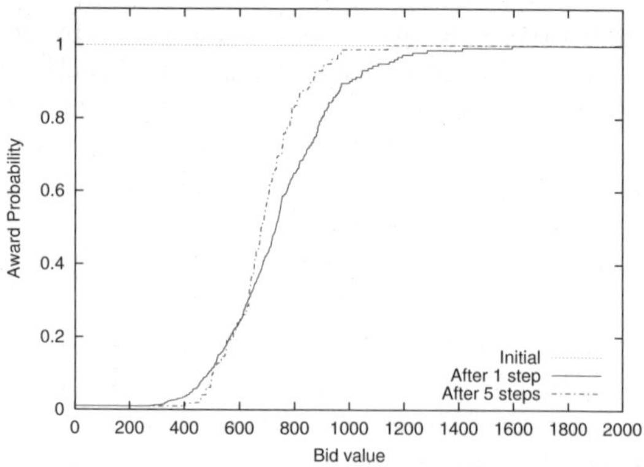

Fig. 2. Award probabilities for RT auctions after several learning steps

of the robots. Given the set $B = WB^{RT} \cup MLB$, the probability of receiving a bid with a value greater than b is computed as:

$$P(bid > b) = \frac{|(x \in B|x > b)|}{|B|}$$

As with the award probabilities, the auctioneer generates a random number, and if it is less than or equal to $P(bid > b)$ it offers the task.

Figure 1 shows this probability for bids from 0 to 2000. As expected, there is a high probability of receiving a bid with a low minimum value, while this probability decreases as the minimum required value increases.

4.3 Iterative learning

The probabilities shown in the previous (and subsequent) figures are learned off-line. That is, we let the system run for a fixed period of time, gather information about auctions, bids and awards and, once the run is finished, we use this information and the equations presented above to compute the award and reception probabilities. Initially, the award and reception probabilities are computed using a configuration where robots bid for all tasks and include all tasks in their auctions. However, the scenario is not the same when the robots actually *use* the probabilities, since there will be fewer tasks being bid for or auctioned. Thus, we repeat the process for several steps, updating the probabilities after each run, in order to learn better probability distributions.

Figure 2 shows how the award probabilities for RT auctions change after 1 and 5 learning steps. In the initial step all the probabilities are set to 1, which is equivalent to including all tasks in auctions and sending all bids. The figure shows that the probability of mid-level bids winning increases, while the probability of low-level bids winning decreases (mainly because fewer of them are offered up for auction). Similar graphs are obtained for OT auctions and for the reception probabilities, which are not shown due to space limitations.

	NP	AuP	AllP
Num Rocks	422.37 (10.14)	424.27 (9.38)	415.0 (9.75)
Num Tasks Auctioned	1394.05 (93.89)	350.45 (31.06)	523.35 (103.01)
No Awards	65.18% (2.32)	52.38% (2.43)	74.24% (3.82)
Messages	13606.31 (443.06)	8608.77 (153.27)	3814.81 (265.68)

Table 1. Results for each configuration. Average values (and standard deviations)

Note that we assume that the experimental environment does not change from run to run. That is, the number and distribution of tasks are similar, and the number of robots in the team remains constant. If the environment were different at each run, the probabilities would be of limited help since the robots would be facing a totally different scenario.

5 Experimental results

We have used the FIRE (Federation of Intelligent Robotic Explorers) multi-robot system [6] to evaluate how well our approach performs when using the learned probabilities. The scenario involves multiple robots exploring the surface of Mars. The goal of the team is to characterize a set of rocks at different (known) locations. A task is described as the location of the rock, the type of sensor to be used, and the reward to be paid upon completion. In our experiments, we used a scenario with 5 robots and only one type of rock, which all the robots can characterize. Each rock has the same reward of 2000.

This scenario has been used in each of the following configurations: *No Probabilities (NP)*, where robots bid for all the tasks being offered and they auction all their remaining tasks, *Auction Probabilities (AuP)*, robots use the reception probabilities to decide whether to include a task in a Call for Bids and bidders bid on all tasks, and *All Probabilities (AllP)*, both award and reception probabilities are used.

Table 1 shows the results obtained for each configuration using the probabilities obtained after 5 learning steps. To evaluate them, we have focused on the following aspects: *number of rocks* characterized by the team (i.e. performance), number of *tasks auctioned* by robots, percentage of auctions for which *no awards* could be made (including those for which there were no bids), and *number of task messages* sent (we consider a message offering a task in a Call for Bids and a bidding message for a task to have the same cost).

Observing the results, we can see that the *Auction Probabilities* configuration performs as well as the *No Probabilities*, while reducing the communication cost by almost 40%. Although we were expecting a small decrease in performance when using the reception probabilities (since not all the tasks would be auctioned), the high frequency of auctions (each robot is constantly trying to auction its tasks) makes that tasks are considered for being offered often enough so that they are eventually auctioned and assigned to a better robot, and therefore the performance is not affected. As for the *All Probabilities* configuration, while it uses only a third of the number of messages used in the *No Probabilities* configuration, its performance degrades somewhat. This drop in performance is due to the fact that since not all robots bid for all tasks, some of the tasks may not be awarded to the optimal robot.

Regarding the number of auctions with no awards, the first thing to point out is that in the *No Probabilities* configuration, there is a high percentage of such auctions. The reason is that when a robot auctions any of its pending tasks, it is sometimes already the best suited robot for that task, and none of the bids sent by other robots is good enough for the task to be traded. However, the robot keeps trying to auction it over and over. Although some of the tasks are awarded to other robots (and this does indeed improve the performance of the system, see [6]), most of these auctions are a waste of communication and computation resources. This percentage is reduced in the *Auction Probabilities*, since only those tasks with high chances of being successfully bid for (and thus, awarded) are offered to the other robots. Although we expected similar results in the *All Probabilities* configuration, we found that the percentage of auctions without awards did not decrease but increased (actually, almost 60% of the auctions received no bids because of the robot's selective bidding). The problem is that the auction and bid probabilities are "acting against each other". On one hand, a robot usually auctions a task if the probability of receiving a successful bid is high. On the other hand, a bidder usually bids for those tasks for which its bid has high probability of being awarded. However, when a robot auctions a task for which a low minimum bid is sufficient, it is often the case that the bids computed by the other robots are also low, therefore with a low probability of being awarded, and usually not sent, and thus, the auctioneer does not receive any bid.

Regarding the computational cost, it can be observed that the *Auction Probabilities* and *All Probabilities* configurations drastically reduce the number of tasks being auctioned by the robots (they are offering only 25% and 37% of the tasks offered by the *No Probabilities* configuration, respectively). This has a major impact on the computation cost incurred by the Scheduler, since very few bids have to be computed.

We have also evaluated how accurate are the learned probabilities. For the award probabilities, each "bid decision" a bidder faced was classified as: bid sent and won, bid sent and lost, bid not sent but would have won, and bid not sent and would have lost. From these four classes, the first and last ones are right decisions, while the other two are wrong decisions. Figure 3 shows the percentage of right decisions as a function of the award probability. As we can see, there is an improvement of at least 30% (up to 60% for some probability values) of the percentage obtained when not using probabilities. Moreover, the right decision percentage is much higher at the ends than in the middle. This make sense, since at both ends the bidder is very confident that it will either lose or win the auction. However, in the mid-range probabilities, it is not clear what is the right thing to do, since the outcome of the decision

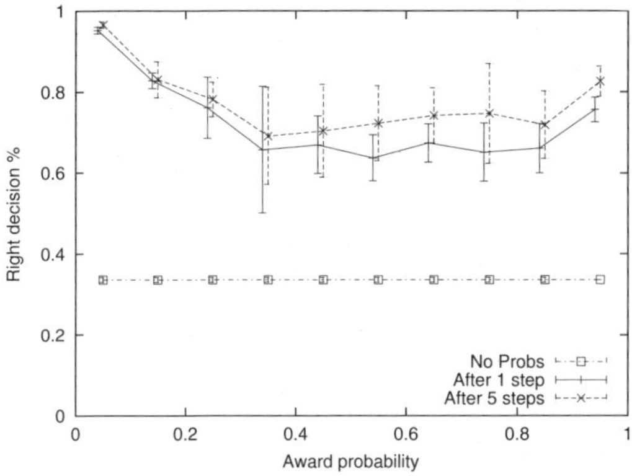

Fig. 3. Right decision percentage for award probability for RT auctions

is almost 50-50. We can also observe that this percentage improves after 5 learning steps. We used a similar method to evaluate the award probability for OT auctions and the reception probability, computing the percentage of tasks awarded over those being offered. The results show that the probabilities also lead to a higher award percentage, with similar improvements. We do not show them due to space limitations

6 Discussion and Future work

Coordinating a team of robots is still a major issue in the field of robotics. One important aspect of the coordination is the multirobot task allocation problem. Many approaches try to solve this problem, one of them being the market-based approach. However, this approach usually has a high cost in communication and computational resources, which has usually not been addressed. In this paper, we have presented a method for reducing these costs.

To do so, the robots learn whether a task is worth bidding for and also whether it is worth offering any of its pending tasks to the other robots. The robot learns two probability functions, the award probability and the reception probability, that are used to decide *when to bid* and *when to auction.*

The results show that the use of probabilities considerably decreases both communication and computation costs. Moreover, the *Auction Probabilities* configuration does not affect the performance at all. Thus, if performance is the main concern, this configuration is the most adequate. However, if the main concern is communication cost, then the *All Probabilities* configuration should be chosen, since it drastically decreases this cost, but it does affect the performance somewhat.

For future work, we would like to investigate if a better model for the award probabilities can be developed, so that its effect on performance can be eliminated. We would also like to extend our probability method so that it can be used when the system works with combinatorial auctions. The difficulty in this case is that the bids sent by the robots are not for single tasks, as in the work we have presented in this paper, but they are for bundles of task. Thus, it is not clear how the bids should be treated when computing the probabilities (we could use either the absolute value, the ratio between bid and reward of the bundle, or some other function).

An assumption we have made is that the robots repeatedly perform their mission in a similar environment, so that the probabilities learned offline can be used in future missions. It would be interesting to have this learning process on-line, so that the robots can learn the probabilities while performing the given mission. This would allow for using the method on one-time missions, and would also allow the robots to adapt to dynamic environments where the distribution of bids and tasks could change while performing their task.

References

1. V. Cicirello and S. Smith. Amplification of Search Performance through Randomization of Heuristics. In *Proceedings of the 5th International Conference on Principles of Constraint Programming (CP 02)*, 2002.
2. M. B. Dias and A. Stentz. A Free Market Architecture for Distributed Control of a Multirobot System. In *Proceedings of the 6th International Conference on Intelligent Autonomous Systems*, pages 115–122, 2000.
3. A. Gage and R. Murphy. Affective Recruitment of Distributed Heterogeneous Agents. In *Proc. of 19th National Conference on AI*, pages 14–19, 2004.
4. B. Gerkey and M. Matarić. Sold! Auction methods for multi-robot control. *IEEE Transactions on Robotics and Automation*, 18(5):758–768, 2002.
5. B. Gerkey and M. Matarić. A formal analysis and taxonomy of task allocation in multi-robot systems. *Int. J. of Robotics Research*, 23(9):939–954, 2004.
6. D. Goldberg, V. Cicirello, M. B. Dias, R. Simmons, S. Smith, and A. Stentz. Market-Based Multi-Robot Planning in a Distributed Layered Architecture. In *Multi-Robot Systems: From Swarms to Intelligent Automata: Proceedings from the 2003 International Workshop on Multi-Robot Systems*, volume 2, pages 27–38. Kluwer Academic Publishers, 2003.
7. M. Golfarelli, D. Maio, and S. Rizzi. A Task-Swap Negotiation Protocol Based on the Contract Net Paradigm. *Tech. Report, 005-97, CSITE (Research Centre for Informatics and Telecommunication Systems), University of Bologna*, 1997.
8. K. Larson and T. Sandholm. Costly Valuation Computation in Auctions. In *Proc. of Theoretical Aspects of Reasoning about Knowledge*, pages 169–182, 2001.
9. P. Milgrom. Auctions and bidding: A primer. *Journal of Economic Perspectives*, 3(3):3–22, 1989.
10. L. Parker. Alliance: An architecture for fault-tolerant multi-robot cooperation. *IEEE Transactions on Robotics and Automation*, 14(2):220–240, 1998.

System Identification of Self-Organizing Robotic Swarms

Nikolaus Correll and Alcherio Martinoli

Swarm-Intelligent Systems Group, École Polytechnique Fédérale Lausanne
firstname.lastname@epfl.ch

We discuss system identification of self-organizing, swarm robotic systems using a "gray-box" approach, based on probabilistic macroscopic models. Using a well known case study concerned with the autonomous inspection of a regular structure by a swarm of miniature robots, we show how to achieve highly accurate predictive models by combining previously developed probabilistic modeling and calibration methods, with parameter optimization based on experimental data (80 experiments involving 5-20 real robots).
Key properties of the optimization process are outlined with the help of a simple scenario and a model that can be solved analytically. Concepts are then validated numerically for the more complex, non-linear inspection scenario.

1 Introduction

Self-organization emerges from the interplay of four ingredients. Positive feedback (e.g., amplification) and negative feedback (e.g., saturation, resource exhaustion), randomness, and multiple interactions between individuals [2]. Although self-organization might achieve less efficient coordination than other distributed control schemes, it can provide extremely high levels of robustness and can be applied to miniature robotic platforms such as those mentioned in this paper.

For designing and formally analyzing self-organized robotic systems, appropriate models are necessary. Modeling allows us to focus on key design parameters, and costly, time-consuming experiments can be reduced to a minimum. As randomness and fully distributed control are at the core of self-organized swarm coordination, we make use of probabilistic macroscopic models to statistically capture the average dynamics and performances of a self-organized robotic swarm.

System modeling and identification [6, 8] is the process of deriving a mathematical model for a metric of interest from observed data. For swarm-robotic systems, probabilistic models have been successfully applied to several case

studies (see for instance [9] or [7, 11]), and lead to good quantitative agreement between reality and the prediction of the model. In this contribution, we take a step forwards and show how the identification process can be complemented by optimization procedures for improving the quantitative agreement between model prediction and experimental data. Such improved models can then serve as a baseline for exploring other aspects of the system without performing additional experiments.
As a case study we consider a homogeneous swarm engaged in the inspection of a regular structure [5], where modeling assumptions (homogeneous distribution of robots and objects, see [9] for more details) are only partially fulfilled. In this case, optimization of model parameters allows us not only to achieve accurate prediction in the model, but also yields valuable insight into improving the structure of the model. Although results are validated on a particular case study, the proposed method is generally applicable for system identification of self-organized robotic systems whose dynamics can be captured by probabilistic models.

2 Probabilistic Modeling of Swarm-Robotic Systems

As more extensively detailed in [9] we abstract the robots' behavior as an arbitrary Probabilistic Finite State Machine (PFSM), whose states are chosen according to the metric of interest. Interactions among the robots or with the environment are represented by state transitions and abstracted to encountering probabilities, whereas the time spent in a certain state is captured by the average interaction time. We hereby assume that the robots and objects are uniformly distributed in the environment, that the system is markovian (i.e. the system's state depends only on its previous state), and that the encountering probabilities scale linearly with the number of objects (i.e. the chance to encounter an object is ten times as high when there are ten objects than when there is only one), which is reasonable when the ratio of free-space to space occupied by robots and objects is large.

Calibration of Model Parameters: Following [4, 9], we calculate the *geometric probability* of encountering an object from the ratio of the object's detection area (the area in that it can be detected by another robot), and the total area of the arena. The (unit less) geometric probability can then be converted into the object's *encountering probability* per time-step, using a simple heuristic based on the area that a robot sweeps with its sensors in this period (based on the characteristics of its sensors and its speed).
The interaction time—if not directly specified within the robot's controller—cannot be calibrated as easily, but needs to be measured using systematic experiments with two robots, or one robot and one object [5].
We validated this approach systematically for simple scenarios in [4], and it showed quantitatively correct predictions for various swarm-robotic case

studies [9]. Nevertheless, the calibration methodology reaches its limitations in overcrowded scenarios, and for nonuniform spatial distributions of robots (as in [5], for instance).

3 Identification of a Linear Swarm-Robotic System

For introducing the concepts applied in this paper more formally, we consider a simple case study first. A swarm of robots is moving on a bounded arena, performing collision avoidance with other robots and the boundary walls using a reactive controller [1]. We are now interested in finding a model for the average number of robots in the search and collision avoidance states from experimental data. Finding a model involves three basic steps [8]:

1. Performing a set of experiments measuring the metric of interest for different parameter choices;
2. Deriving a candidate model;
3. Finding optimal parameters that minimize the mismatch between candidate model and experimental data.

Note, that we consider the system in discrete time, as the observed data for system identification is collected by sampling.

An Identification Experiment: For simplicity, we consider a system without time varying inputs, where an experiment is characterized over a time interval $0 \leq k \leq n$ by its state vector $N(k)$ and parameters that are set by the experimenter (e.g., the number of robots N_0). The state variables are measurements of an arbitrary metric of interest, for instance, the average number of robots searching at time k, $N_s(k)$.

A Candidate Model: Following the methodology outlined in section 2, we model the system by a PFSM with three states: search, avoidance of walls, and avoidance of robots. Our approach involves the following assumptions. Every time step, one of N_0 robots can encounter another robot with probability p_r (and any other robot with probability $p_R = p_r(N_0 - 1)$), and a wall with probability p_w. We further assume that a collision can be sufficiently characterized by its mean duration (T_r and T_w). We can then write the following set of difference equations:

$$\underbrace{\begin{pmatrix} N_{ar}(k+1) \\ N_{aw}(k+1) \\ N_s(k+1) \end{pmatrix}}_{N(k)} = \underbrace{\begin{pmatrix} 1-\frac{1}{T_r} & 0 & p_R \\ 0 & 1-\frac{1}{T_w} & p_w \\ \frac{1}{T_r} & \frac{1}{T_w} & 1-p_r-p_w \end{pmatrix}}_{\theta} \begin{pmatrix} N_{ar}(k) \\ N_{aw}(k) \\ N_s(k) \end{pmatrix}, \quad (1)$$

and the initial conditions

$$\left(N_{ar}(0)\ N_{aw}(0)\ N_s(0) \right)^T = (0 \quad 0 \quad N_0)^T \quad (2)$$

with $N_s(k)$ being the number of robots searching at time k, $N_{ar}(k)$ the number of robots avoiding a robot, $N_{aw}(k)$ the number of robots avoiding a wall, and

N_0 the total number of robots. We can interpret the first row of (1) as follows. The number of robots avoiding another robot is decreased by those that return from a collision ($\frac{1}{T_r}N_{ar}(k)$), and increased by searching robots colliding with another robot ($p_R N_s(k)$). The other rows of (1) can be interpreted in a similar fashion. Equation (1) can be reformulated as

$$\hat{N}(k+1)^T = N(k)^T \theta^T, \tag{3}$$

where $\hat{N}(k+1)$ is the *estimate* based on the measurements of the real system $N(k)$ and the parameters θ.

Analytical Optimization: Provided the state vector measurements $N(k) = (N_{ar}(k) \quad N_{aw}(k) \quad N_s(k))^T$ in the interval $0 \leq k \leq n$, we can now calculate the prediction error of the model estimate $\hat{N}(k)$. Optimal parameters (θ) can then be found using the least-squares method that can be formulated as

$$\min_\theta \frac{1}{n} \sum_{k=1}^{n} (N(k)^T - \hat{N}(k)^T)^2 = \min_\theta \frac{1}{n} \sum_{k=1}^{n} (N(k)^T - N(k-1)^T \theta^T)^2 \tag{4}$$

We denote the matrix θ that minimizes (4) by $\hat{\theta}_n$:

$$\hat{\theta}_n = \arg\min_\theta \frac{1}{n} \sum_{k=1}^{N} (N(k)^T - N(k-1)^T \theta^T)^2 \tag{5}$$

Since (4) is quadratic in θ, we can find the minimum value easily by setting the derivative of (4) to zero:

$$0 = \frac{2}{n} \sum_{k=1}^{n} N(k-1)(N(k)^T - N(k-1)^T \hat{\theta}_n^T), \tag{6}$$

yielding

$$\hat{\theta}_n = \left[\sum_{k=1}^{n} N(k-1) N(k-1)^T \right]^{-1} \sum_{k=1}^{n} N(k-1) N(k)^T, \tag{7}$$

which is straightforward to compute given the availability of the measured state variables $N(k)$.

Initial Parameter Estimation: In the above experiment, measurements for all state variables (N_{ar}, N_{aw}, and N_s) are available. Imagine now that it is not possible to measure N_{ar} and N_{aw} independently from each other (this is reasonable for collisions with robots close to the wall for instance). Then, p_R and p_w cannot be estimated (1), but only $1 - p_R - p_w$ (using numerical methods). As a work-around, additional experimental data need to be gathered by varying other parameters, for instance changing the number of robots. Such a procedure leads to an identification problem with a smaller number of degrees of freedom, but it might not be feasible to conduct it for every single parameter; in particular for systems where the ratio of parameters to the number of observed state variables is high. Then, an initial estimate using the calibration heuristic from [9] is extremely helpful, as shown below.

4 Identification of a Non-Linear Swarm-Robotic System

We consider a case study introduced in [4, 5]. A swarm of miniature robots (*Alice* [3]) is concerned with the inspection of a simplified jet turbine (Figure 1, *right*). Robots are randomly searching through a bounded arena, and inspect objects in the environment (turbine blades) by circumnavigating them. The robots's controllers are endowed with a timeout (T_{max}), that indicates when a blade can be left. After the timeout is expired, the robot leaves the blade at the next tip it encounters (compare the Finite State Machine (FSM) of the individual robot in Figure 1, *middle*).

We have selected this case study for the following reasons: in [5], our mod-

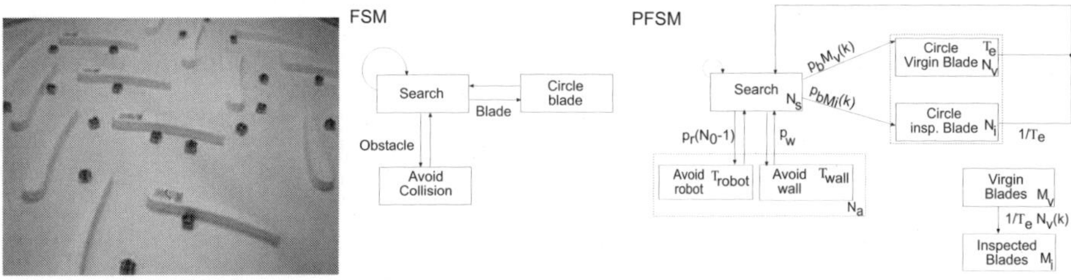

Fig. 1. *Left*: Experimental setup involving 16 blades and up to 20 miniature robots. *Middle*: Finite State Machine describing individual robot's behavior. *Right*: Probabilistic Finite State Machine describing the state of an individual robot and the environment (number of blades inspected).

eling approach showed certain discrepancies when compared to the dynamics of the real system. We identified two problems. First, the regular structure of the environment imposing drift on the robotic swarm, which in turn favors the inspection of some parts of the arena due to nonuniform distribution of the robots. Second, noisy perception of environmental features led to artifacts such as robots getting stuck at the outer walls or at the tips of individual blades. Using system identification, we show how to get further insight into the system, that the model is indeed able to reproduce the observed dynamics, and that it has the potential to serve as a predictor for the system.

Identification Experiment: The simplified turbine environment was implemented in a rectangular arena of $1.10m \times 1m$ [5]. The robots are endowed with a PIC micro controller (368 bytes RAM, 8Kb FLASH), have a length of 22mm, and a maximal speed of 4cm/s. Four IR modules can serve as crude proximity sensors (up to 3cm) and local communication devices (up to 6 cm). We performed four sets of 20 runs each (80 experiments altogether) for $N_0 = 5, 10, 16$ and 20 robots in an environment with $M_0 = 16$ blades. The inspection progress (the number of inspected blades) is monitored (sampling rate 1Hz) using an overhead camera and the tracking software *SwisTrack* [1], that records the robots's trajectories at a frame rate of 30Hz.

[1] http://swistrack.sourceforge.net

Candidate Model: The model for the inspection case study is extensively detailed in [4, 5], and will only briefly be summarized here. Following our modeling methodology [9], we sketch a PFSM of the system dynamics according to our metric of interest (the number of blades inspected), based on an individual robot's FSM (compare Figure 1, middle and right). Initially, a robot is in search mode. At every time-step, a robot may encounter the outer wall, an other robot, or a blade with probabilities p_w, p_R, and p_b, respectively, causing it to enter an avoidance or inspection state. Every state is associated with an interaction time (T_w, T_r, and T_b); the average time the robot spends in this state. In the probabilistic model, this is equivalent to leaving the state with probability $\frac{1}{T_w}$, $\frac{1}{T_r}$, and $\frac{1}{T_b}$.
In our PFSM we also keep track of whether the robot is inspecting a "virgin" blade or a previously inspected blade. With N_0 being the number of robots, and M_0 the number of blades, we can hence derive the following difference equations for the robotic states:

$$\hat{N}_{ar}(k+1) = \hat{N}_{ar}(k) + p_r\hat{N}_s(k)(N_0 - 1) - \frac{1}{T_r}\hat{N}_{ar}(k) \tag{8}$$

$$\hat{N}_{aw}(k+1) = \hat{N}_{aw}(k) + p_w\hat{N}_s(k) - \frac{1}{T_w}\hat{N}_{aw}(k) \tag{9}$$

$$\hat{N}_v(k+1) = \hat{N}_v(k) + p_b\hat{M}_v(k)\hat{N}_s(k) - \frac{1}{T_b}\hat{N}_v(k) \tag{10}$$

$$\hat{N}_i(k+1) = \hat{N}_i(k) + p_b\hat{M}_i(k)\hat{N}_s(k) - \frac{1}{T_b}\hat{N}_i(k) \tag{11}$$

$$\hat{N}_s(k+1) = N_0 - \hat{N}_{ar}(k+1) - \hat{N}_{aw}(k+1) - \hat{N}_v(k+1) - \hat{N}_i(k+1) \tag{12}$$

and for the environmental states:

$$\hat{M}_v(k+1) = \hat{M}_v(k) - \frac{1}{T_b}\hat{N}_v(k) \tag{13}$$

$$\hat{M}_i(k+1) = M_0 - \hat{M}_v(k+1) \tag{14}$$

with initial conditions $N_s(0) = N_0$ and $M_v(0) = M_0$, and all other states zero. Note, that the system described above is non-linear and thus cannot be written in matrix notation as the simple model in section 3.

Initial Calibration of Model Parameters: During the experiments we recorded only the inspection progress itself ($M_i(k)$), whereas the number of unknown parameters is 6 (encountering probabilities and interaction times for blades, robots, and walls). Although the problem of an under-determined system as outlined in section 3 is mitigated by performing experiments with different numbers of robots, there is still an infinite number of possible combinations for encountering probabilities and interaction times. For example, inspection of one blade could be very fast, whereas collisions take very long, or the other way around; both leading to the same measured inspection time. As this is acceptable so long as the system dynamics are reproduced faithfully, the resulting models might perform poorly when used as predictors for other parameters or modified robot controllers.

In order to improve the potential predictive quality of the model after identification, we provide the identification process with an initial guess using the calibration heuristic from [9], which we have good reason to believe [4] yields values that come close to the parameters of the "real" system. Initial guesses for the parameters of the experiment are picked up from [5], and are summarized in Table 1.

Parameter Optimization: We are interested in finding the optimal parameter set θ_N that minimizes the difference between the model's prediction $\hat{M}_v(k, \theta, N_0)$ and measurements of the real system $M_v(k, N_0)$. At the same time, we want to minimize the difference between the optimal parameters and the initial guess $\theta_0 = (p_r \quad p_w \quad p_b \quad T_r \quad T_w \quad T_b)$ (compare Table 1). We thus formulate the following optimization problem

$$\theta_n = \arg\min_{\theta} \left[\left(\sum_{N_0 \epsilon R} \sum_{k=0}^{n} (\hat{M}_v(k, \theta, N_0) - M_v(k, N_0))^2 \right) + (\theta - \theta_0)^2 \right] \tag{15}$$

With $R = \{5, 10, 16, 20\}$ the number of robots in an experiment (N_0), and n the duration of the longest experiment in the set. Obtaining an analytical solution to (15) is unfeasible. Nevertheless, as it is quadratic in θ, we can attempt to solve it using a convex optimization algorithm (for instance *fmincon* in Matlab^{TM}). Here, the optimization routine integrates (14) numerically until $\hat{M}_v(k) = 0.01$ ($\hat{M}_v(k)$ converges to zero asymptotically) to find an optimal solution for θ.

Table 1. Initial guesses for model parameters (heuristic calibration/measurements of the real system), time discretization of the model T=1s

	Wall	Blade	Robot
Encountering probability	$p_w = 0.045$	$p_b = 0.0106$	$p_r = 0.0015$
Interaction time	$T_w = 10.00$s	$T_{hb} = 10.00$s	$T_r = 4.00$s

5 Results

The numerical optimization routine finds an optimal parameter set for minimizing model prediction error with respect to experimental measurements after 300 to 600 function evaluations (each involving the calculation of model prediction for four different team sizes). We used values from Table 1 as initial guess for θ, and provided upper and lower bounds on θ (all encountering probabilities are forced to be in the interval between 0.00001 and 1, whereas the interaction times are bounded by $8 \leq T_b \leq 12$, $2 \leq T_r \leq 20$, and $2 \leq T_w \leq 30$). Optimal parameters (θ_n) are given in Table 2. Inspection

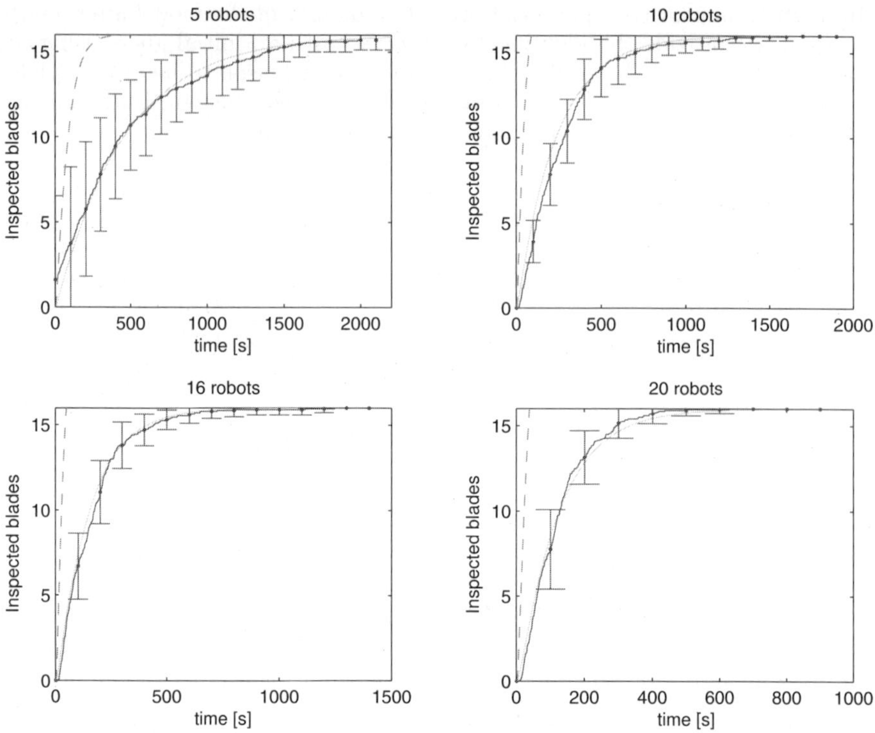

Fig. 2. Average inspection progress for different team sizes and model predictions before (dashed curve) and after optimization (continuous curve) when compared to experimental data (continuous curve with error bars). Error bars show the standard deviation of the measurements.

progress $M_v(k)$ (mean and standard deviation), as well as the model estimates before ($\hat{M_v}(k, N_0, \theta_0)$) and after optimization ($\hat{M_v}(k, N_0, \theta_n$) for different team sizes ($N_0 = 5, 10, 16, 20$) are given in Figure 2.

Table 2. Model parameters after optimization, time discretization of the model T=1s

	Wall	Blade	Robot
Encountering probability	$p_w = 0.18899$	$p_b = 0.001543$	$p_r = 0.009363$
Interaction time	$T_w = 11.16556s$	$T_{hb} = 8.0666s$	$T_r = 3.829s$

6 Discussion

Model prediction based on the initial estimate θ_0, predicted the inspection to be roughly twice as fast as reality [5]. Using optimization procedures to complement the system identification process allowed for a good match between prediction and experimental data, for the inspection time metric as well as for the inspection progress. As a consistent pattern over all runs (with different initial conditions and parameter bounds), one observes that the encountering probability for detecting a blade is estimated to be an order of magnitude lower than the resulting probability after calibration. This result matches our expectations from [5] where we observed a nonuniform distribution of the robots over the arena (induced by the geometry of the blades and the robots's controllers), which yields a blade encountering probability that is dependent on the location in the arena.

This case study is an example where optimization averages out more complex effects, such as the drift phenomenon, and allows us to capture them with a relatively simple model. In order to reach this level of accuracy in a different way, the level of detail in our model would need to be increased at the cost of analytical tractability.

As formally shown in section 3, a system where not all of the state variables are measured is likely to be under-determined if the number of degrees of freedom are not varied separately. Indeed, although the predictions for the robot-to-robot encountering probability are consistent over all runs (the system was run with different numbers of robots), it is unclear how the optimization procedure should distribute robots between the wall avoidance state, and the inspection state in order to achieve the desired inspection delay. This is reasonable, as the differences between model prediction and real experiments can not only be explained by the robots' spatial distribution, but also by the fact that the robots may get stuck on a wall or prematurely leave blades. These artifacts are not explicitly captured by the model considered here, and therefore need to be accounted elsewhere in the model's structure. In order to reduce the number of degrees of freedom in the system and increase identifiability [6], one could either measure more state variables, or perform other sets of experiments (in our case, possibly an experiment in a larger arena or with more blades).

7 Conclusion

We have shown how our methodology for system identification of swarm robotic systems, consisting of deriving the system dynamics from the controller of the individual robot, and calibration of model parameters using simple heuristics, can be complemented by macroscopic data-fitting with parameter constraints. We have also shown how this additional step can not only lead to quantitatively correct models for systems that do not comply with the assumptions of our calibration procedure, but also give additional insight into

the whole system. We note that the proposed methodology is suitable for a wide range of swarm robotic scenarios, in which dynamics can be captured by probabilistic models.
Concerning our case study swarm robotic inspection, optimization results show that the structure of the model is sufficient to accurately reproduce the system dynamics, and indicates critical points where the model reaches its limitations. Indeed, the probability of encountering a blade, which is an order of magnitude lower than the calibration estimate, suggests further research into capturing the spatio-temporal distribution of the robots, potentially by using diffusion models borrowed from statistical physics [10].

Acknowledgments

Both authors are sponsored by a Swiss NSF grant (contract Nr. PP002-68647).

References

1. R. Arkin. *Behavior-Based Robotics.* The MIT press, Cambridge, MA, USA, 2000.
2. E. Bonabeau, M. Dorigo, and G. Theraulaz. *Swarm Intelligence: From Natural to Artificial Systems.* SFI Studies in the Science of Complexity, Oxford University Press, New York, NY, USA, 1999.
3. G. Caprari and R. Siegwart. Mobile micro-robots ready to use: Alice. In *Proceedings of IEEE/RSJ International Conference on Intelligent Robots and Systems*, pages 3295–3300, 2005.
4. N. Correll and A. Martinoli. Modeling and optimization of a swarm-intelligent inspection system. In *Proceedings of the 7th Symposium on Distributed Autonomous Robotic System (DARS).* To appear in Springer Distributed Autonomous Systems VII, 2004.
5. N. Correll and A. Martinoli. Collective inspection of regular structures using a swarm of miniature robots. In *Proceedings of the 9th International Symposium of Experimental Robotics (ISER)*, pages 375–385. Springer Tracts for Advanced Robotics (STAR), Vol. 21., 2006.
6. Rolf Johansson. *System Modeling and Identification.* Prentice Hall, 1993.
7. K. Lerman and A. Galstyan. Mathematical model of foraging in a group of robots: Effect of interference. *Autonomous Robots*, 2(13):127–141, 2002.
8. Lennart Ljung. *System Identification — Theory for the User.* Prentice Hall, 1999.
9. A. Martinoli, K. Easton, and W. Agassounon. Modeling of swarm robotic systems: A case study in collaborative distributed manipulation. *Int. Journal of Robotics Research*, 23(4):415–436, 2004.
10. A. Prorok. Multi-level modeling of swarm robotic inspection. Master's thesis, SWIS-MP1, École Polytechnique Fédérale Lausanne, 2006.
11. K. Sugawara, M. Sano, I. Yoshihara, and K. Abe. Cooperative behavior of interacting robots. *Artificial Life and Robotics*, 2:62–67.

Synchronization Control by Structural Modification of Nonlinear Oscillator Network

Tetsuro Funato, Daisuke Kurabayashi, and Masahito Nara

Tokyo Institute of Technology, Ookayama 2-12-1, Meguro-ku, Tokyo 152-8552, Japan, {funato,dkura,m-nara}@irs.ctrl.titech.ac.jp

Summary. The structural features of a system significantly affect the attributes and functions of the system. The effect of this phenomenon can be widely observed, from areas such as the WWW to the brains of animals. In the present paper, a method for controlling the behavior of a system by manipulating the structure is examined using a coupled nonlinear oscillator model. We first describe a property of the eigenfrequencies of coupled oscillators and show that convergent transition is possible by connecting oscillators with significantly different eigenfrequencies. Moreover, using the eigenvalues of a graph matrix, we reveal that a combination of distant oscillators can shift the converged state independent of the eigenfrequencies.

Key words: Network Structure, Nonlinear Oscillator, Synchronization

1 Introduction

Morphological approaches to biological motion represented by a passive dynamic walker[1] are attracting attention particularly with regard to the relationship between morphological properties and functional creation. Faculties derived from topology as such are not restricted only to motion. In the field of networks, The small world (SW) structure[2, 3] has a high transferring efficiency, and networks with an approximately power-law vertex degree distribution, such as the Internet, strongly resist the random removal of nodes[4]. These types of generation phenomena can also be seen in the brain. Synapses, the connections between neurons, are classified into electrical and chemical types according to their transmission mechanisms. Chemical transmission synapses are mediated by message-carrying chemicals, and the combination of chemicals can be changed by the amount of message-carrying chemicals. This transformation is used for memory and learning abilities. For instance, the tendency of crickets to change their behavior based on past experience is caused by the variation of message-carrying chemicals influenced by nitric oxide (NO)[5].

In order to understand the relationship between structure and function, an appropriate model is necessary. Nonlinear oscillators can serve this purpose because they can generate the forces for synchronization. Research regarding the nonlinear oscillator has been carried out for a long period. For example, Linkens[6, 7] analyzed a coupled van der Pol (VDP) oscillator system, particularly with respect to its convergence. Among recent studies, Kuramoto[8] proposed a model that has a phase as only one free parameter and indicated that synchronization is affected by the connection coefficient. Jadbabaie et al.[9] and Earl et al.[10] investigated the relationship between network structure and convergence by introducing geometric factors to the Kuramoto model.

Current oscillator models can be effectively used as a brain model; this is based on some biochemistry reports, such as one indicating that the synchronization of neurons is essential in the visual information processing of a cat[11]. In particular, oscillatory neural networks (ONNs) proposed by employing the oscillatory feature to a neural network (NN), the most popular information processing model of the brain, has attracted considerable attention. As an example, Hoppensteadt[12] realized associative memory with synchronization by introducing a Hebbian learning rule to the Kuramoto model.

Using ONNs, research on the structural features of the oscillator network may facilitate a discussion on the morphology of the brain. However, in order to consider the changes in synapses, the examination of a change in the synapse structure cannot be avoided. Nonetheless, there is little research on convergence in this regard, and current research mainly considers mainly the static properties. The purpose of this research is to elucidate the dynamic phenomenon of structural change and to control convergence by the manipulation of only the transition of structure.

We first understand a property of the phase gap of coupled oscillators using the Kuramoto model and propose a control method for the convergence state. Next, we confirm this method via simulation. Then, by using the eigenvalues of a graph matrix, we describe a geometric manipulation that enables a converged transition independent of the eigenfrequencies.

2 Models and Characteristics of Oscillators

2.1 Nonlinear oscillator

In order to consider the relationship between the system and structure, an appropriate model is required. In this study, we constructed a model by considering the characteristics of nonlinear oscillators as follows:

1. Connected nonlinear oscillators can cause the forces required for synchronization.
2. The converged states can be changed by these forces.

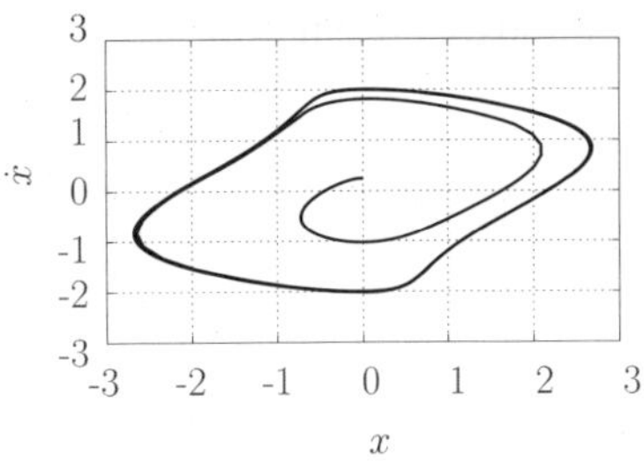

Fig. 1. Limit Cycle

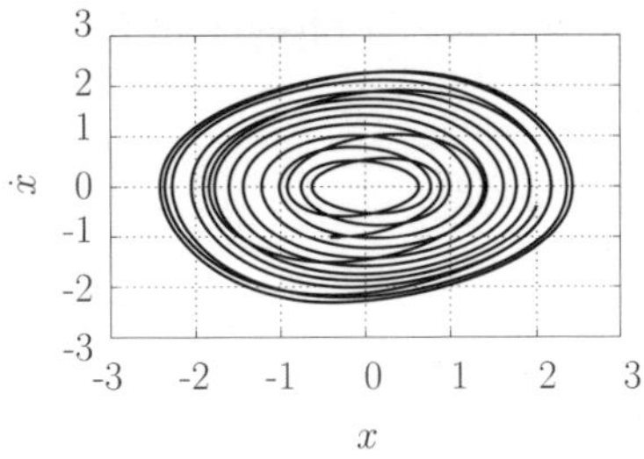

Fig. 2. Quasi-Periodic

The transition of converged states considered in this study is shown by the apparent change from the cyclic state (limit cycle) shown in Fig. 1 to the non-cyclic state (quasi-periodic oscillation) shown in Fig. 2.

By using nonlinear oscillators as models, a structural change produces forces between the oscillators in the system(property 1), and this effect can change the state of the oscillators, i.e., the state of the system (property 2).

2.2 Kuramoto model

We used the Kuramoto model[8] as nonlinear oscillators for this model. In this model, the phase of the ith oscillator θ_i is as follows:

$$\dot{\theta}_i = \omega_i + \frac{K}{N}\sum_{j=1}^{N} \Gamma(\theta_j - \theta_i). \tag{1}$$

Here, K is the connection coefficient; ω_i, the eigenfrequency, Γ, the interaction function; and N, the total number of oscillators. We suppose that Γ is odd-symmetric ($\Gamma(\theta) = -\Gamma(-\theta)$) and $\Gamma(0) = 0$.

3 Convergent Transition via Properties of Eigenfrequency

3.1 Characteristics of coupled oscillators

We express a group of oscillators connected to the ith oscillator as O_i and that connected to the jth oscillator as O_j. When we consider further connections between the ith and jth oscillators, the transition can be represented by a change in γ (0→1); the equations of motion regarding these oscillators then become

$$\dot{\theta}_i = \omega_i + \frac{K}{N+\gamma}\left\{\sum_{s\in O_i} \Gamma(\theta_s - \theta_i) + \gamma\Gamma(\theta_j - \theta_i)\right\} \tag{2}$$

$$\dot{\theta}_j = \omega_j + \frac{K}{N+\gamma}\left\{\sum_{t\in O_j} \Gamma(\theta_t - \theta_j) + \gamma\Gamma(\theta_i - \theta_j)\right\}. \tag{3}$$

In this case, a phase gap between the ith and jth oscillators $\phi(=\theta_i-\theta_j)$ is expressed as

$$\dot{\phi}=\dot{\theta}_i-\dot{\theta}_j=\omega_i-\omega_j+\frac{K}{N+\gamma}\left\{\sum_{s\in O_i}\Gamma(\theta_s-\theta_i)-\sum_{t\in O_j}\Gamma(\theta_t-\theta_j)\right\}-\frac{2K\gamma}{N+\gamma}\Gamma(\phi) \tag{4}$$

$$\frac{\partial\phi(t,\gamma)}{\partial t}=\delta\omega+\frac{K}{N+\gamma}\left\{\sum_{s\in O_i}\Gamma(\theta_s-\theta_i)-\sum_{t\in O_j}\Gamma(\theta_t-\theta_j)\right\}-\frac{2K\gamma}{N+\gamma}\Gamma\left(\phi(t,\gamma)\right). \tag{5}$$

In the case widthout any connection,

$$\frac{\partial\phi(t,0)}{\partial t}=\delta\omega+\frac{K}{N}\left\{\sum_{s\in O_i}\Gamma(\theta_s-\theta_i)-\sum_{t\in O_j}\Gamma(\theta_t-\theta_j)\right\}. \tag{6}$$

When N is sufficiently large, by the approximation of $\frac{K}{N+\gamma}\simeq\frac{K}{N}\left(1-\frac{\gamma}{N}\right)=\frac{K}{N}-\frac{\gamma}{N^2}\simeq\frac{K}{N}$,

$$\begin{aligned}\frac{\partial\phi(t,\gamma)}{\partial t}&\simeq\delta\omega+\frac{K}{N}\left\{\sum_{s\in O_i}\Gamma(\theta_s-\theta_i)-\sum_{t\in O_j}\Gamma(\theta_t-\theta_j)\right\}-\frac{2K\gamma}{N}\Gamma\left(\phi(t,\gamma)\right)\\&=\frac{\partial\phi(t,0)}{\partial t}-\frac{2K\gamma}{N}\Gamma\left(\phi(t,\gamma)\right).\end{aligned} \tag{7}$$

This equation indicates that the addition of new combinations produces an effect of adding (or substituting) an integer multiple of the interaction function.

If the interaction function $\Gamma(\phi(t,\gamma))=\sin\phi(t,\gamma)$,

$$\frac{\partial\phi(t,\gamma)}{\partial t}=-\frac{2K\gamma}{N}\sin\left(\phi(t,\gamma)\right)+\frac{\partial\phi(t,0)}{\partial t}, \tag{8}$$

which denotes a sine function with center $\frac{\partial\phi(t,0)}{\partial t}$ and amplitude $\frac{2K\gamma}{N}$.

From eq.8, following properties can be determined:

1. If $\left|\frac{\partial\phi(t,0)}{\partial t}\right|<\frac{2K\gamma}{N}$, there exists a certain ϕ that satisfies $\left|\frac{\partial\phi(t,0)}{\partial t}\right|=0$.
2. If $\left|\frac{\partial\phi(t,0)}{\partial t}\right|>\frac{2K\gamma}{N}$, no ϕ satisfies $\left|\frac{\partial\phi(t,0)}{\partial t}\right|=0$.

Similar to the concept of frequency locking[13], converged states become a limit cycle in case 1 because of the existence of an equivalent point and become a quasi-periodic oscillation in case 2. Figs. 3 and 4 display the graphs of eq.8 under the condition that $\frac{\partial\phi(t,0)}{\partial t}$ is constant.

As a result, we can control the converged states from the limit cycle to the quasi-periodic oscillation by coupling oscillators whose $\left|\frac{\partial\phi(t,0)}{\partial t}\right|$ is large, i.e., the gap in eigenfrequencies is large.

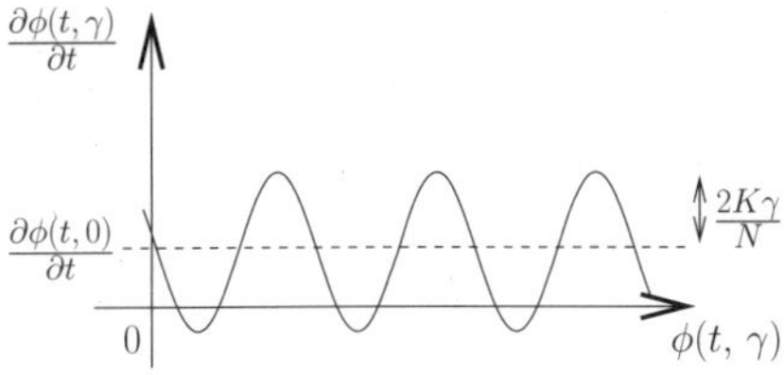

Fig. 3. $|\frac{\partial\phi(t,0)}{\partial t}| < \frac{2K\gamma}{N}$

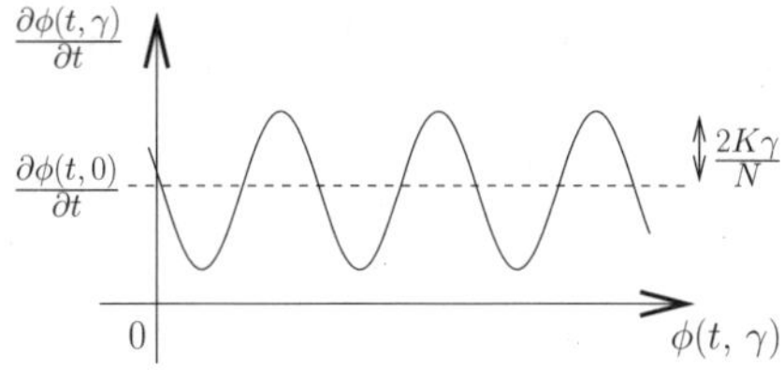

Fig. 4. $|\frac{\partial\phi(t,0)}{\partial t}| > \frac{2K\gamma}{N}$

Fig. 5a. Structural transition

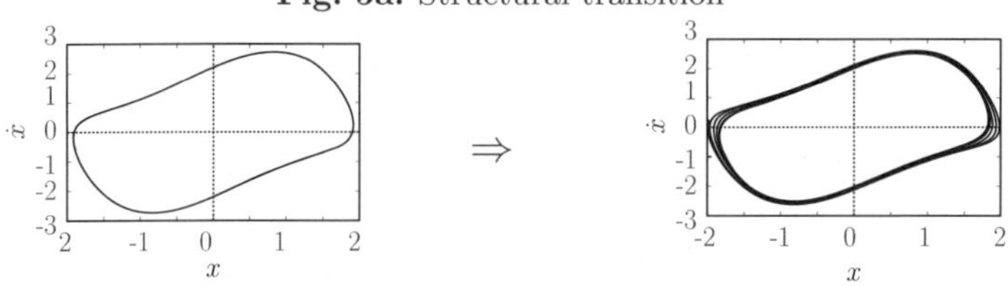

Fig. 5b. Convergent shift by structural transition

3.2 Simulation

We simulated the state of the oscillators by altering the connection structure based on the eigenfrequencies (Fig. 5a and 5b). We used VDP oscillators in this simulation. The equation of a VDP oscillator is $\ddot{x_i} - \epsilon_i(1 - x_i^2)\dot{x_i} + \omega_i^2 x_i = 0$, and the interaction forces for synchronization is expressed as $x_i(t+1) = x_i(t) + K\left\{\frac{1}{N_i(t)}\sum_{j=1}^{N_i(t)} x_j(t) - x_i(t)\right\}$. Here, $x_j(t)$ is the state of an oscillator that is connected to the ith oscillator and $N_i(t)$ is the total number of the connected oscillators.

In Fig. 5a, the vertices of the graph represent the oscillators, the edges express the relationship of the connections and the values of the vertices represent the eigenfrequencies.

We change the structure by creating a new connection between oscillators whose eigenfrequencies are significantly different. Through the simulation, converged states shift from the limit cycle to quasi-periodic oscillations; this is confirmed by Fig. 5b.

4 Convergent Transition via only Geometric Properties

4.1 Kuramoto model considering structural disposition

Jadbabaie et al.[9] proposed an oscillator model that builds on the Kuramoto model and includes connection relationship between oscillators.

$$\dot{\theta} = \omega - \frac{K}{N} B \sin\left(B^T \theta\right) \tag{9}$$

The $N \times e$ matrix B represents an oriented graph that has N vertices and e edges, and the following conditions hold:

- If edge j incoming to vertex i, $B_{ij} = 1$.
- If edge j outcoming from vertex i, $B_{ij} = -1$.
- If edge j and i are not connected, $B_{ij} = 0$.

In eq.9, θ and ω are N vectors expressing the phase and eigenfrequency, respectively.

4.2 Convergent condition via eigenfrequency

In addition, Jadbabaie showed that there is at least one convergent oscillator if the connection coefficient K satisfies

$$K > \left(\frac{\pi}{2}\right)^2 \frac{N\lambda_{\max}(L)}{\lambda_{\min}(L)^2} \|\omega\|_2. \tag{10}$$

Here, $L = BB^T$ is a matrix called the Laplacian and $\lambda(L)$ is the eigenvalue of L. $\lambda_{\max}$ and $\lambda_{\min}$ are the maximum and minimum eigenvalues, respectively.

In eq.10, the geometric property is expressed only by

$$\frac{\lambda_{\max}(L)}{\lambda_{\min}(L)^2}, \tag{11}$$

therefore, convergent control is possible by considering the eigenvalues of the Laplacian.

Moreover, a similar conclusion can be gathered from the theses of Pecora et al.[14] and Barahona et al.[15]. They also modeled oscillator states using a geometric matrix and showed that oscillators can converge when the difference between the maximum eigenvalue and minimum eigenvalue goes below a certain constant value.

4.3 Eigenvalues of the Laplacian

In order to investigate the elemental property of structural transition, we consider connecting a pair of oscillators to a simple cycle-graph-shaped oscillator network.

$$B = \begin{bmatrix} -1 & 0 & & \cdots & 0 & 1 \\ 1 & -1 & 0 & \cdots & 0 & 0 \\ & 1 & -1 & & & \\ & & \ddots & \ddots & & \vdots \\ & 0 & & 1 & -1 & 0 \\ & & & & 1 & -1 \end{bmatrix}, \quad L = BB^T = \begin{bmatrix} 2 & -1 & 0 & \cdots & 0 & -1 \\ -1 & 2 & -1 & & & 0 \\ 0 & -1 & 2 & \ddots & & \vdots \\ \vdots & & \ddots & \ddots & & 0 \\ 0 & & & & 2 & -1 \\ -1 & 0 & \cdots & 0 & -1 & 2 \end{bmatrix} \tag{12}$$

We express the matrix of a graph that has an additional combination as $B' = [B|x]$ by using N vector x; $x = \begin{bmatrix} 0 & \cdots & 0 & -1 & \overbrace{0 \cdots 0}^{m} & 1 & \cdots & 0 \end{bmatrix}^T$. In this vector, m denotes the interval of the coupling oscillators. Because the current graph is ring-shaped, the variation in the starting point of the connection makes no difference. Therefore, x can be written as $x = \begin{bmatrix} 0 & -1 & \overbrace{0 \cdots 0}^{m} & 1 & \cdots & 0 \end{bmatrix}^T$ without a lack of generality. Then, the Laplacian of B' becomes

$$L' = BB^T + xx^T = \begin{bmatrix} 2 & -1 & 0 & \cdots & & 0 & 0 & \cdots & & 0 & -1 \\ -1 & 3 & -1 & & & \vdots & -1 & & & & 0 \\ 0 & -1 & 2 & \ddots & & & 0 & & & & \vdots \\ \vdots & & \ddots & \ddots & & 0 & \vdots & & & & 0 \\ & & & & & -1 & 0 & & & & \\ 0 & \cdots & & 0 & -1 & 2 & -1 & 0 & \cdots & & 0 \\ 0 & -1 & 0 & \cdots & 0 & -1 & 3 & -1 & & & \vdots \\ \vdots & & & & & 0 & -1 & 2 & \ddots & & \\ & & & & & & & \ddots & \ddots & & 0 \\ 0 & & & & & \vdots & & & -1 & 2 & -1 \\ -1 & 0 & \cdots & & 0 & 0 & \cdots & & 0 & -1 & 2 \end{bmatrix}. \tag{13}$$

$$|L' - \lambda I| = \begin{array}{c} \begin{matrix} 1 & 2 & 3 & \cdots & m-1 & m & m+1 & \cdots & N-1 & N \end{matrix} \\ \left| \begin{matrix} -1 & 0 & & & \cdots & & & 0 & -1 & 2-\lambda \\ 2-\lambda & -1 & 0 & \cdots & 0 & 0 & \cdots & & 0 & -1 \\ -1 & 3-\lambda & -1 & & \vdots & -1 & & & & 0 \\ 0 & -1 & 2-\lambda & \ddots & & 0 & & & & \vdots \\ \vdots & & \ddots & \ddots & 0 & \vdots & & & & 0 \\ & & & & -1 & 0 & & & & \\ 0 & \cdots & p & 0 \quad -1 & 2-\lambda & -1 & 0 & \cdots & & 0 \\ 0 & -1 & 0 & \cdots \quad 0 & -1 & 3-\lambda & -1 & & & \vdots \\ \vdots & & & & 0 & -1 & 2-\lambda & \ddots & & \\ & & & & & & \ddots & \ddots & & 0 \\ 0 & & & \cdots & & & 0 & -1 & 2-\lambda & -1 \end{matrix} \right| \end{array}. \tag{14}$$

We set $(m+1)\times(m+1)$ matrix S, $(m+1)\times(N-m-1)$ matrix T, $(N-m-1)\times(m+1)$ matrix U and $(N-m-1)\times(N-m-1)$ matrix V as follows:

$$
\begin{array}{cccccccc}
\mathbf{1} & \mathbf{2} & \mathbf{3} & \cdots & & \boldsymbol{m-1} & \boldsymbol{m} & \boldsymbol{m+1}
\end{array}
$$

$$
S=\begin{bmatrix}
-1 & 0 & & & & \cdots & 0 & 0 \\
2-\lambda & -1 & 0 & & & & 0 & \\
-1 & 3-\lambda & -1 & 0 & \cdots & 0 & -1 & \vdots \\
0 & -1 & 2-\lambda & \ddots & & & 0 & \\
\vdots & & \ddots & \ddots & & & \vdots & \\
 & & & & & -1 & 0 & \\
0 & \cdots & & 0 & -1 & 2-\lambda & -1 & 0 \\
0 & -1 & 0 & \cdots & 0 & -1 & 3-\lambda & -1
\end{bmatrix},\quad
T=\begin{bmatrix}
0 & \cdots & 0 & -1 & 2-\lambda \\
 & & & & -1 \\
 & & \ddots & & 0 \\
 & 0 & & & \vdots \\
 & & & & 0
\end{bmatrix},
$$

$$
U=\begin{bmatrix}
0 & \cdots & 0 & -1 & 2-\lambda \\
 & & & & -1 \\
 & & \ddots & & 0 \\
 & 0 & & & \vdots \\
 & & & & 0
\end{bmatrix}\quad
V=\begin{bmatrix}
-1 & & & & & \\
2-\lambda & -1 & & & 0 & \\
-1 & \ddots & \ddots & & & \\
0 & & & & & \\
\vdots & & & & & \\
0 & \cdots & 0 & -1 & 2-\lambda & -1
\end{bmatrix}.
$$

Based on these matrices, eq.14 becomes $|L'-\lambda I| = \left|\begin{smallmatrix} S & T \\ U & V \end{smallmatrix}\right| = |V|\,|S-TV^{-1}U|$. Because $|V|=1$, $|L'-\lambda I|=0 \leftrightarrow |S-TV^{-1}U|=0$.

If we set $V^{-1}=\{v_{ij}\}$ (i, j indicates row and column number respectively),

$$
\begin{cases}
[TV^{-1}U]_{1(N-1)} & = v_{(N-1)1}-(2-\lambda)v_{N1} \\
[TV^{-1}U]_{1N} & = -(2-\lambda)v_{(N-1)1}+(2-\lambda)^2 v_{N1} \\
 & \quad +v_{(N-1)2}-(2-\lambda)v_{N2} \\
[TV^{-1}U]_{2N-1} & = v_{N1} \\
[TV^{-1}U]_{2N} & = -(2-\lambda)v_{N1}+v_{N2} \\
[TV^{-1}U]_{other} & = 0
\end{cases}. \tag{15}
$$

Further,

$$
\begin{bmatrix} -1 & 0 \\ 2-\lambda & -1 \end{bmatrix}^{-1} = \begin{bmatrix} -1 & 0 \\ -(2-\lambda) & -1 \end{bmatrix}
$$

$$
\begin{bmatrix} -1 & 0 & 0 \\ 2-\lambda & -1 & 0 \\ -1 & 2-\lambda & -1 \end{bmatrix}^{-1} = \begin{bmatrix} -1 & 0 & 0 \\ 2-\lambda & -1 & 0 \\ -(2-\lambda)^2+1 & -(2-\lambda) & -1 \end{bmatrix}
$$

$$
\vdots
$$

indicates the following about V^{-1}:

$$
\left[v_{1(N-1)}\ v_{2(N-1)}\ \cdots\ v_{N(N-1)}\right]^T = \left[0\ \cdots\ 0\ -1\ -(2-\lambda)\right]^T \tag{16}
$$

$$
\left[v_{1N}\ v_{2N}\ \cdots\ v_{NN}\right]^T = \left[0\ \cdots\ 0\ -1\right]^T, \tag{17}
$$

therefore,

$$TV^{-1}U = 0. \tag{18}$$

As a result, $|L' - \lambda I| = 0 \leftrightarrow |S| = 0$.

$$|S| = \left| \begin{array}{cccccccc} \mathbf{1} & \mathbf{2} & \mathbf{3} & \cdots & & \boldsymbol{m-1} & \boldsymbol{m} & \boldsymbol{m+1} \\ -1 & 0 & & & & \cdots & 0 & 0 \\ 2-\lambda & -1 & 0 & & & & 0 & \\ -1 & 3-\lambda & -1 & 0 & \cdots & 0 & -1 & \vdots \\ 0 & -1 & 2-\lambda & \ddots & & & 0 & \\ \vdots & & \ddots & \ddots & & & \vdots & \\ & & & & & -1 & 0 & \\ 0 & \cdots & & 0 & -1 & 2-\lambda & -1 & 0 \\ 0 & -1 & 0 & \cdots & 0 & -1 & 3-\lambda & -1 \end{array} \right| = 0 \tag{19}$$

$$\left| \begin{array}{cccc} \mathbf{1} & \mathbf{2} & \cdots & \boldsymbol{m-3} \\ 2-\lambda & -1 & & 0 \\ -1 & 2-\lambda & \ddots & \\ & \ddots & \ddots & -1 \\ 0 & & -1 & 2-\lambda \end{array} \right| = (-1)^m. \tag{20}$$

The size of this determinant is decided by the interval of the coupling oscillators. This result suggests that an even more distant connection induces a larger variation of the eigenvalues. Therefore, $\frac{\lambda_{\max}(L)}{\lambda_{\min}(L)^2}$ increases due to the remote oscillator connection, which breaks the condition of eq.10, and the oscillators become quasi-periodic.

Moreover, the right-hand side of eq.20 is -1 to the power of m. Therefore, the sign changes depending on whether the distance of the oscillators is an even or odd number; this causes a non-trivial difference in the convergent condition.

5 Conclusion

In this study, we examined the effect of structural transition on the behavior of systems and investigated a method whereby convergence can be controlled only by structural manipulation, using an oscillator network as the system model.

We first analyzed the behavior of a network from the viewpoint of phase gap and showed that it is possible to control the converged state by connecting oscillators with significantly different frequencies. In addition, we confirmed this phenomenon by simulation.

We then calculated the rate of maximum and minimum eigenvalues of a graph matrix to find a control method mainly via the structural properties based on the thesis by Jadbabie. Assuming that the oscillator network is constructed on a circular graph, we showed that the variation of eigenvalues depended on the distance of the additional coupled oscillators, i.e., a remote connection can induce convergent transition.

Acknowledgement

This research was supported in part by a grant-in-aid for scientific research on priority areas from the Japanese Ministry of Education, Culture, Sports, Science and Technology (17075007).

References

1. T.McGeer. Passive dynamic walking. *The International Journal of Robotics Research*, Vol. 9, No. 2, pp. 62–82, 1990.
2. D.J.Watts and S.H.Strogatz. Collective dynamics of 'small-world' networks. *nature*, Vol. 393, pp. 440–442, 1998.
3. S.H.Strogatz. Exploring complex networks. *nature*, Vol. 410, pp. 268–276, 2001.
4. D.S.Callaway, M.E.J.Newman, S.H.Strogatz, and D.J.Watts. Network robustness and fragility: Percolation on random graphs. *Physical Review Letters*, Vol. 85, No. 25, pp. 5468–5471, 2000.
5. H.Aonuma, M.Iwasaki, and K.Niwa. Role of no signaling in switching mechanisms in the nervous system of insect. *SICE Proc. 2004*, pp. 2477–2482, 2004.
6. D.A.Linkens. Analytical solution of large numbers of mutually coupled nearly sinusoidal oscillators. *IEEE Trans. on Circuits and Systems*, Vol. CAS-21, No. 2, pp. 294–300, 1974.
7. D.A.Linkens. Stability of entrainment conditions for a particular form of mutually coupled van der pol oscillators. *IEEE Trans. on Circuits and Systems*, Vol. CAS-23, No. 2, pp. 113–121, 1976.
8. Y.Kuramoto. *Chemical Oscillations, waves, and Turbulence.* Springer, 1984.
9. A.Jadbabaie, N.Motee, and M.Barahona. On the stability of the kuramoto model of coupled nonlinear oscillators. In *the American Control Conference*, 2004.
10. M.G.Earl and S.H.Strogatz. Synchronization in oscillator networks with delayed coupling: A stability criterion. *Physical Review E*, Vol. 67, p. 036204, 2003.
11. C.M.Gray, P.Konig, A.K.Engel, and W.Singer. Oscillatory responses in cat visual cortex exhibit inter-columnar synchronization which reflects global stimulus properties. *nature*, Vol. 338, pp. 334–337, 1989.
12. F.C.Hoppensteadt and E.M.Izhikevich. Oscillatory neurocomputers with dynamic connectivity. *Physical Review Letters*, Vol. 82, No. 14, pp. 2383–2386, 1999.
13. H.Haken. *Advanced Synergetics.* Springer, 1983.
14. L.M.Pecora and T.L.Carroll. Master stability functions for synchronized coupled systems. *Physical Review Letters*, Vol. 80, No. 10, p. 2109, 1998.
15. M.Barahona and L.M.Pecora. Synchronization in small-world systems. *Physical Review Letters*, Vol. 89, No. 5, pp. 054101–1 – 054101–4, 2002.

Frontier-Graph Exploration for Multi-robot Systems in an Unknown Indoor Environment

Mark Gossage, Ai Peng New and Chee Kong Cheng

Cooperative Systems and Machine Intelligence Lab, DSO National Laboratories, Singapore {gmark, naipeng, ccheekon}@dso.org.sg

We present a new method for single/multiple robot indoor exploration and mapping. The algorithm combines local Frontier-based exploration technique and global graph-based representation of the environment to produce a robust autonomous exploration strategy. This graph is used and shared to allow cooperative exploration. Our implementation is fully decentralised and has no central control to organise the robots, it is also robust to failures both in communications and robot attrition. Our approach has been demonstrated to work on a team of two Pioneer 3AT robots in an area of $50m^2$. In the simulator it has been successfully scaled to a team of five robots in a map of over a hundred rooms and an area of $5000m^2$.

1 Introduction

The aim of this research work is to develop a robust exploration technique for single and multi-robot system in an unknown indoor environment. For the case where multiple robots are employed, the system must be scalable, decentralised, tolerant to temporary lost of communication between some robots and able to handle robot attrition.

Our approach uses concepts from the Frontier-based exploration and graph-based representation of the environment. Instead of using a global occupancy grid (OG), our method uses a fixed size local OG centred on each robot. As the robot explores, the frontiers located in the OG are used as graph nodes on a global graph. The robot uses this global graph for high-level planning and navigation. For multi-robot system, the graph is regularly synchronised between robots to ensure a consistent map. Cooperative exploration is performed through tasks negotiated between robots.

Section 2 discusses other related work in single robot indoor exploration and multi-robot systems. Section 3 describes our Frontier-Graph Exploration algorithm (FGE) in detail. In section 4, the simulation and hardware testing results are presented and discussed. Finally in section 5, further work is considered.

2 Related Work

Exploration of an environment is a complicated problem for robots to solve. It consists of many sub-problems, many of which are non-trivial to solve. Problems such as Simultaneous Localisation and Mapping (SLAM), determining where to explore next, deciding how best to organise multiple platforms have all enjoyed substantial attention over the years. From our survey of exploration algorithms, there are only a few map representation methods. The majority are based upon grid based map representation (Burgard et al. 2002; Zlot et al. 2002, Simmons et al. 2000, Yamauchi 1998) with a few others looking at topological map representations (Choset and Burdick 1995a, 1995b).

By far the most common method for multi robot exploration is to use an occupancy grid (Moravec 1998) with a frontier based exploration algorithm (Yamauchi 1996). It is a simple and proven method of exploration and mapping. But the grid based map has a fundamental flaw associated with it, its size. Grid maps (especially those of high resolution) require large amounts of memory (usually in the order of megabytes). With the growth rate of computer memory, this does not necessarily present a problem. However in a multi-robot context, such maps must be shared regularly. Bandwidth consumption becomes very high as the entire map is needed for exploration and path planning.

The other method of map representation is topological (graph based) such as the Generalised Voronoi Graph (GVG) (Choset and Burdick 1995a, 1995b). Though the graph representation is simple and compact, the GVG is a technique primarily for exploring tunnel-like environments or corridors. In a cluttered environment (such as an office), it becomes difficult to create the graph, and the graph complexity increases greatly.

This paper presents a hybrid exploration and mapping algorithm, called 'Frontier-Graph Exploration' (FGE). It attempts to use the best parts of both grid based maps and topological maps while mitigating the various limitations of each representation.

The idea of developing a hybrid grid-topological representation is not new. Thrun and Bücken (Thrun and Bücken 1996) showed a method converting a grid based map into a region based (topological) map. More recently the Centibot project (Konolige et al. 2004) used such a map in its Spatial Reasoning component. After the initial mapping (using an occupancy grid), an offline process was applied to turn this into a Voronoi diagram, and to the final graph. The graph was then used by the robots for sharing and navigation about the map.

Our method distinguishes itself from these past works in that it is generated online as the robot explores, not as a post-processing step after the exploration is performed. It also does not require the global occupancy grid that the other methods require, but uses a local grid only. This substantially reduced bandwidth usage, as no grid information is ever shared, only the graph information. It is also a decentralized approach with robots individually creating and sharing graphs as they explore.

3 Frontier-Graph Exploration

3.1 Single Robot Frontier-Graph Exploration

The FGE algorithm assumes a 2D environment, a robot mounted with an accurate sensor (such as a Laser Range Finder), and good localisation.

The main perception component of the FGE is the 'Circular Perception'; a small fixed size local OG (Elfes 1990) formed by accumulating sensor data as the robot moves in the environment. It is used to determine obstacles, configuration space and frontiers for exploration. Unlike a normal occupancy grid, which will continually expand as the robot moves, the Circular Perception is a fixed size grid that scrolls about with the robot, always keeping the robot in the middle of the grid. This perception provides sufficient information for local sensing and navigation, but consumes a fixed amount of memory.

The basic FGE algorithm can be summarised as: find the frontiers on the local OG, add them to the graph, select a frontier (using a simple cost function) and move to it, repeat until no frontiers exist. A step by step diagram of the exploration can be seen in figs. 1a-c.

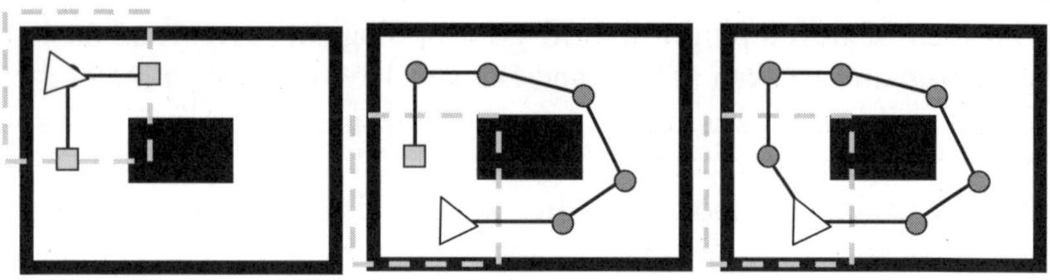

(a) (b) (c)

Fig. 1a-c. Step by step frontier-graph exploration. The triangle represents the robot, dashed lines are the circular perception, squares are unexplored nodes, circles are explored nodes

The first problem with this approach was how to determine when the robot moves into space that has already been explored. In Frontier exploration, this is taken care of by the global OG. But a local OG does not hold this history. Therefore when the robot returns to an explored space, it may add new frontiers in the explored area. For example in the fig. 1b, the robot has clearly completed the exploration. However without the global knowledge, it might continue adding new nodes to the graph and would continue to explore indefinitely.

In order to overcome this problem, the FGE algorithm stores and makes use of the clear space region information. The clear space region is defined as a polygon approximating the clear (non-configuration space) area that is around the robot. Each explored node in the graph includes the clear space region information.

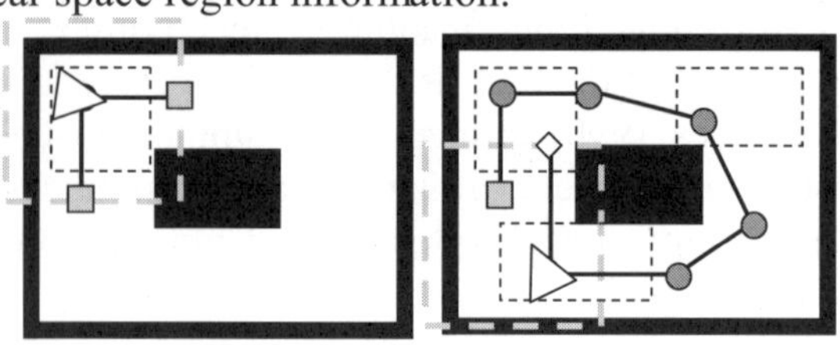

(a) (b)

Fig. 2a-b. FGE with added region information (dotted line, not all regions are added for clarity). The diamond represents a possible frontier that will not be added into the graph as it falls within an existing region

In figs. 2a & b above, the robot has performed FGE and added region information into the graph. As it nears the completion of the exploration, it reaches back to an explored area (fig. 2b). The Circular Perception reports a frontier at the point marked with a diamond. However this point is within an existing clear space region (the region attached to the first node), therefore the frontier should not be added to the graph.

The second problem was that of creating too many frontier nodes. As the robot finds frontiers, they are added to the graph. In normal frontier exploration, as the robot moves to a frontier, other nearby frontiers will be removed. However within FGE, graph nodes must still be explored to ensure completeness. This causes the robot to perform useless work, exploring nodes that will not yield new information.

Using the region information, it is possible to identify these 'extra work' nodes. Upon arriving at a node, all the unexplored nodes, which are in the current clear space region, but not near to a frontier, can be considered 'unnecessary to explore'. In fig. 2b when the robot reaches the final node, the current region information, overlaps the last remaining unexplored node (the square). Since it is clear that the node is in the explored space and not near any frontiers, it can be marked as 'unnecessary to explore'.

To summarise, the full Frontier-Graph Exploration Algorithm is as follows:

1. Move forward a distance to perform initial population of circular perception.
2. Add a graph node (type: explored) at the robot's current location as the starting point, with the clear space region attached.
3. For each node within the current clear space region:
   - Add link between the node and the current node (unless present).
   - If the node is an unexplored node and is not within a certain distance of a frontier, mark the node as unnecessary to explore.
4. For each of the frontiers found within the circular perception:
   - If the frontier is within an existing region (excluding the current region) do not add this.
   - Else add this frontier as an unexplored node and add an edge linking the current node to this unexplored node.
5. If there are unexplored nodes, select an unexplored node (according to the cost function) and move the robot to the node. Otherwise end.
6. Upon arriving at the node, change that node's type from unexplored to explored.
7. Goto Step 3.

3.2 Multi-Robot Frontier-Graph Exploration

Building on the foundation of single robot FGE, we extended the algorithm to multiple robots. The two main features to enable multi-robot exploration are graph merging, to maintain a consistent global graph, and node ownership/negotiation mechanism.

In Multi-Robot FGE, we assumed the robots have good localisation and a common frame of reference, but not necessarily the same start position. We also assumed that both robots and communication could fail. Within the simulator, the localisation assumption is acceptable. Within the hardware however, this is not so easy to achieve.

In order for each robot to maintain a consistent global graph, the graph was regularly broadcast to all other robots. As the global graph is quite compact (even with the region information), it is feasible to send the entire graph in one go. Because of this, robots could maintain a consistent graph unless they were out of communication range for some time.

The individual robots do not attempt to maintain a common numbering scheme for its nodes and edges, but instead use the location of the nodes to merge the graph. This requires the common frame of reference, but is relatively simple to do. When a robot received a graph update, it would match each node in turn with its existing nodes using the nodes location (with a small distance tolerance), and then add any new nodes to its graph. At the same time, it would build a temporary mapping between the received graph node ids and its own graph node ids. Then the edges would be added, using the mapping to determine which nodes should be linked. In practice this was found be acceptable, and performed well in simulation and hardware, provided the localisation was consistent.

The node ownership/negotiation mechanism is used to help the robots coordinate among themselves. Each node is assigned an owner (the robot who discovers the node). When a robot needs a task to do, it takes all unexplored nodes in the graph and sorts them into a list using a cost function based upon the distance between the robot and the node (using Dijkstra's shortest path).

If the lowest cost node is owned by the robot, it performs exploration as normal. If it does not own the node, it must contact the owner and negotiate with them for node ownership. If a robot is unable to obtain the node, it will then consider its second choice and so on. Should no nodes be available, the robot will wait for new ones to appear.

When a robot is contacted with a negotiation request it will agree to transfer the node if the robot is not planning to do it. If it is moving to the node, then it would compare its distance to node with the requesting robot, and agree if the requester is nearer the node.

If the requesting robot could not contact the node owner, it would assume that the owner has either failed, or is too far away from the node to explore it. In which case, it would 'steal' node ownership. Though this assumption is not always correct, it allow the overall algorithm to be robust to communication/platform failures.

4 Results and Discussion

We performed both single robot and multiple robot exploration (only multi-robot results presented) in both the Player/Stage simulation and on our Pioneer 3AT hardware.

4.1 Multiple Robot Simulation

Multi-robot simulation was conducted in the hospital environment as shown in fig. 3. The hospital map is the largest map available with Player/Stage simulator. Measuring around 120m x 50m, and with over a hundred rooms, it is a formidable test for robotic exploration. A team of five simulated robots managed to explore the map in around forty five minutes, generating a graph of several hundred nodes in the process. During our tests, one of the robots failed. However the remaining robots were able to cope with this loss and complete the task without issue.

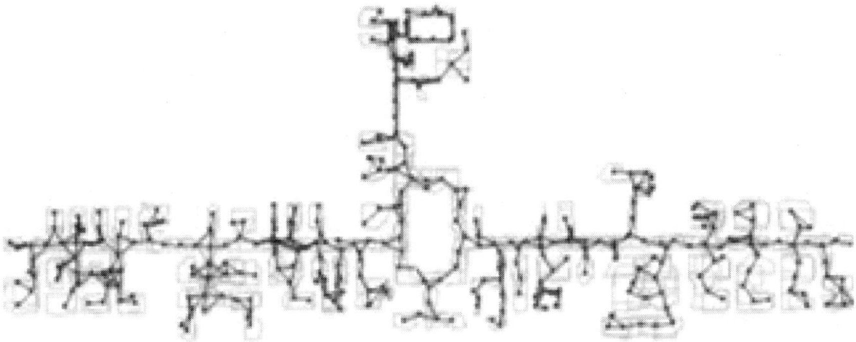

Fig. 3. Multi-Robot FGE in a simulated environment of 120m x 50m, showing the graph nodes and the regions.

4.2 Multiple Robot Hardware

The FGE algorithm was tested in hardware using two Pioneer 3AT's in our company's lobby area. It is consists of a relatively open area, and is about $50m^2$ in size.

One of the biggest issues during hardware testing was localisation. We used a decentralised EKF based SLAM algorithm to provide localisation. It was mentioned earlier that FGE required a common frame of reference for graph sharing and merging. However errors in the localisation often caused the robots graphs to become misaligned with each other. This would cause robots to be unable to navigate to the tasks that they negotiated from other robots.

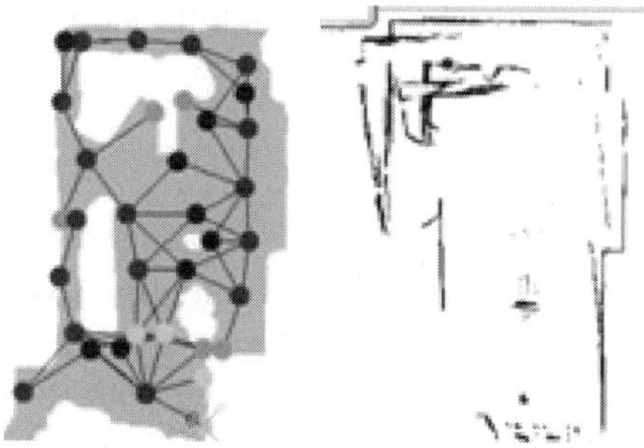

(a) (b)

Fig. 4a-b. Multi-robot FGE in a 50m^2 lobby area, showing the graph and regions (a), and the point cloud map (b)

Fig. 4a-b shows the results from two robots performing FGE. Looking at Fig. 4b, it is clear that some localisation drift has occurred which caused the echoing effect in the point cloud.

Attempts to expand beyond two robots proved impossible as the localisation errors became worse when tests were run with more robots. More work is needed in the localisation area to rectify this problem. However the result here serves to demonstrate that the FGE can be realised in the hardware.

4.3 Issues

Besides localisation, there are two other issues observed: inefficiency in the cooperation and occasional missed frontiers. Often robots stopped to wait for new node to appear or crowded together 'fighting' for nodes. This is generally caused by the simple greedy heuristic used. This could be overcome by using a better cooperation strategy.

The missing frontiers issue is shown in Fig. 5. Here a frontier is on the right of the perception, but due to the narrowness of the configuration space obstacle, the FGE is unable to place a node near to the frontier and it is missed. This limitation causes some frontiers which are beyond narrow openings to be missed. A possible solution to this might be to use some kind of local path planner, though how best to represent this on the graph is still not clear.

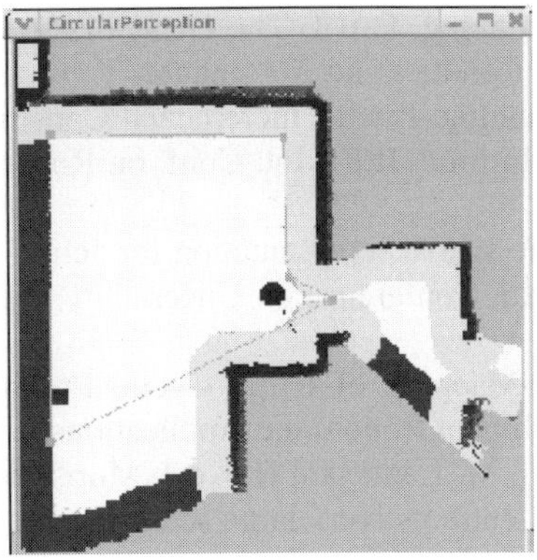

Fig. 5. Due to narrow opening to the right of the robot, it misses the frontier to the right of the robot

5 Further work

The FGE algorithm has several area's for future work. One area is to use an alternative method for task selection and cooperation methods, using ideas from some of earlier mentioned papers (Burgard et al 2002, Simmons et al 2000, Zlot et al 2002).

Localisation is a key area for improvement upon. Several authors (Lu and Milios 1998, Gutman and Konolige 1999) consider a map to be a graph like structure with laser range finder scans attached to each node. This is similar in structure to the FGE, though the purpose is for mapping/localisation. Producing an integrated exploration, mapping and localisation algorithm looks to being a very promising area forward.

The final area for exploration is that of node placement. The current FGE algorithm places its nodes at the limits of its perception, making it a non-repeatable method. An alternative would be to consider using a method closer to the Voronoi Diagram, such as in (Konolige et al. 2004). This would provide a more repeatable method for placement of nodes. A comparison of these methods is an area for future research.

References

Burgard W, Moors M, Schneider F (2002) Collaborative Exploration of Unknown Environment with Teams of Mobile Robots. Lecture notes in Computer Science, volume 2466 pp 52-70

Choset H, Burdick J (1995) Sensor based planning, Part I: The Generalized Voronoi Graph. In Proc. IEEE Int. Conf. on Robotics and Automation

Choset H, Burdick J (1995) Sensor based planning, Part II: Incremental Construction of the Generalized Voronoi Graph. In Proc. IEEE Int. Conf. on Robotics and Automation

Elfes A (1990) Occupancy grids: A stochastic spatial representation for active robot perception. In Proceedings of the Sixth Conference on Uncertainty in AI. Morgan Kaufmann Publishers inc.

Gutman J, Konolige K (1999) Incremental Mapping of Large Cyclic Environments. Proceeding of Conference on Intelligent Robots and Applications

Konolige K, Fox D, Ortiz C, Agno A, Eriksen M, Limketkai B, Ko J, Morisset B, Schulz D, Stewart B, Vincent R (2004) Centibots: very large scale distributed robotic teams. In Proceedings of the International Symposium on Experimental Robotics

Lu F, Milios E (1998) Robot pose estimation in unknown environments by matching 2d range scans. Journal of Intelligent and Robotic Systems

Moravec H (1998) Sensor fusion in certainty grids for mobile robots. AI magazine, pp 61-74

Simmons R, Apfelbaum D, Burgard W, Fox D, Thrun S, Younes H (2000) Coordination for multi-robot exploration and mapping. In Proceedings of the National Conference on Artificial Intelligence. AAAI

Thrun S, Bücken A (1996) Integrating Grid-Based and Topological Maps for Mobile Robot Navigation. In Proceedings of the Thirteenth National Conference on Artificial Intelligence

Yamauchi B (1996) A frontier-based approach for autonomous exploration. In Proceedings of the Thirteenth National Conference on Artificial Intelligence

Yamauchi B (1998) Frontier-based exploration using multiple robots. In Second International Conference on Autonomous Agents, pp 47-53

Zlot R, Stentz A, Dias M B, Thayer S (2002) Multi-Robot Exploration Controlled By A Market Economy. Proceedings of the IEEE International Conference on Robotics and Automation (ICRA)

Player/Stage website http://playerstage.sourceforge.net/

Distributed Robotic: a Language Approach

Claude Guéganno[1] and Dominique Duhaut[2]

[1] University of South Britanny – France claude.gueganno@univ-ubs.fr
[2] University of South Britanny – France Dominique.Duhaut@univ-ubs.fr

Summary.

In this paper we present a powerful and versatile architecture dedicated for robotic and mechatronic systems. The originalities of our study are, (i) that we consider a robot with a dynamic and explicit language approach and (ii) that the communication aspects are abstracted and take place as a natural part in the language. This approach allows easy transfers towards other fields of research like network of sensors, ambient intelligence and ubiquitous robotic. These works concern low-cost micro-system easy to embed in little mechatronic devices. For demonstration of the effectiveness of our architecture and developing tools, we have implemented it in the `maam` robot which is a reconfigurable robot composed of several modules autonomous for CPU, energy and motion. Some results can be found in the last part.

1 Introduction

Distributed architecture for control are widely studied ([1],[2],[3]) but generally address system fitted with operating system and network functionalities. The promising field of ambient intelligence with its variation in ambient and ubiquitous robotic has similar requirements but with reduced capacities. In the field of collective robotic, swarms or teams of robots or even in reconfigurable robotic with autonomous units, some of these requirements are: to distribute the roles for a mission, to take unforeseen events into account, to localize neighbors, to report status to the host. The behavior of the group must be reactive inside a planification itself function of time. Two examples of collaboration are shown in figure 1. In the first one, the robots are associating to make a chain in order to move together, in the second one, the chain is moving, but some agents are left as radio-relays in order to keep a data link with the host. So, the agents must be able to collaborate autonomously to achieve the mission or a part of it. In some cases, the robots must work without any external help (host unreachable). Obviously, the communication aspect is very important in such systems. To take full requirements into account we generally need line in sight communication for solving localization and docking problems and, on an upper layer a network communication for full control purpose including request of reachable robots, broadcast of new plans, control/command of subsets of robots ...

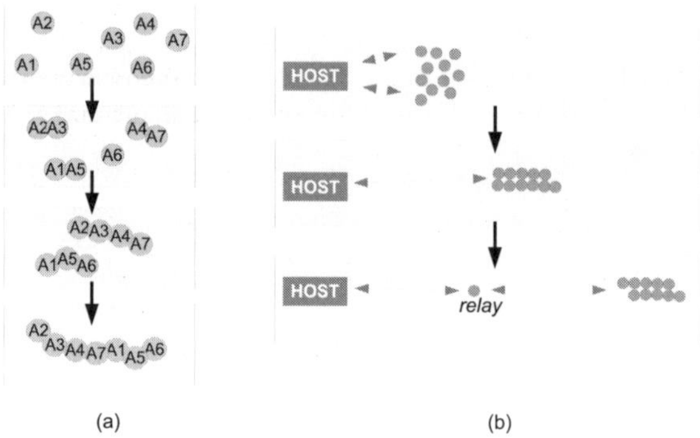

Fig. 1. Two kinds of collaboration. (a)The agents are building a chain. (b)They scatter some of them to ensure a communication between the host and all of them.

So we propose an architecture which gives answers to the following constraints:

- taking into account a centralized control system;
- distributing the intelligence between agents;
- giving possible autonomy for each agent, or subsets of agents;
- allowing collaboration between agents (autonomous group);
- reacting at events by possibly broadcasting new plans towards agents or subset of agents;
- each agent may control a group and in the same time be controlled by another entity.

Moreover, this architecture must not decide of the kind of control which will be apply to the society of agent. It could be deliberative or reactive, or one after the other.

Our proposition is that each agent is likened to a local language, so, controlling it involves (i) to request its own language (or, at least, its set of instructions) (ii) to generate and upload programs according to its particular capacities, and/or to remote-control it directly.

The next section presents the embedded part of this architecture.

2 An architecture for ambient and distributed robotic

2.1 External overview

As we have seen, managing a team of robots with effectiveness suppose to guess the actual robots reachable in the area, to learn the capabilities of each one, to be able to distribute an application between them, and, possibly to remote-control any of them.

The figure 2 summarizes the functionalities of our architecture, and show how we can use it from a control point of view. The first important aspect is that the control system is not supposed to know who are the robots in its environment. It has to inquire for them, an to achieve a mission, to compose with the robots which are actually presents and ready to use (Fig 2-a). Since robots could be different (family, or version in a family) the host requests for the identity of each of them (Fig 2-b). This is done by downloading an `XML` description from the robots. After a parsing stage, the host can generate dedicated user-friendly tools for each agent for remote control (Fig 2-c) and also programming tools (Fig 2-d). Indeed, the `XML` file permits to reconstitute the particular language of the robots, and also gives informations about remote invocations allowed towards the robot. So new programs can be uploaded in the agents (Fig 2-e). A new program replaces the current one instantaneously. All the previous operation from inquiry to uploading can be chained in a standalone program (Fig 2-f).

2.2 Internal overview

Embedded architecture

The embedded architecture is built around a configurable system on chip including a FPGA based hardware easy to adapt for many mechatronic and robotic applications and a micro-controller. It communicates with an industrial *bluetooth* module driven at the Host Control Interface level. More informations about the hardware can be found in [4]

The software include (i) a communication manager made of two finite state automates, one dedicated for permanent inquiry of neighbors, and the other for general communication purpose (remote control, up/download of files); (ii) an event scheduler that can launch some particular instructions of the main program; (iii) an interpreter which runs the main program of the agent (figure 3-left). The IO functionalities of the robot are directly accessed trough the communication interface as well as from the interpreted program (figure 3-right).

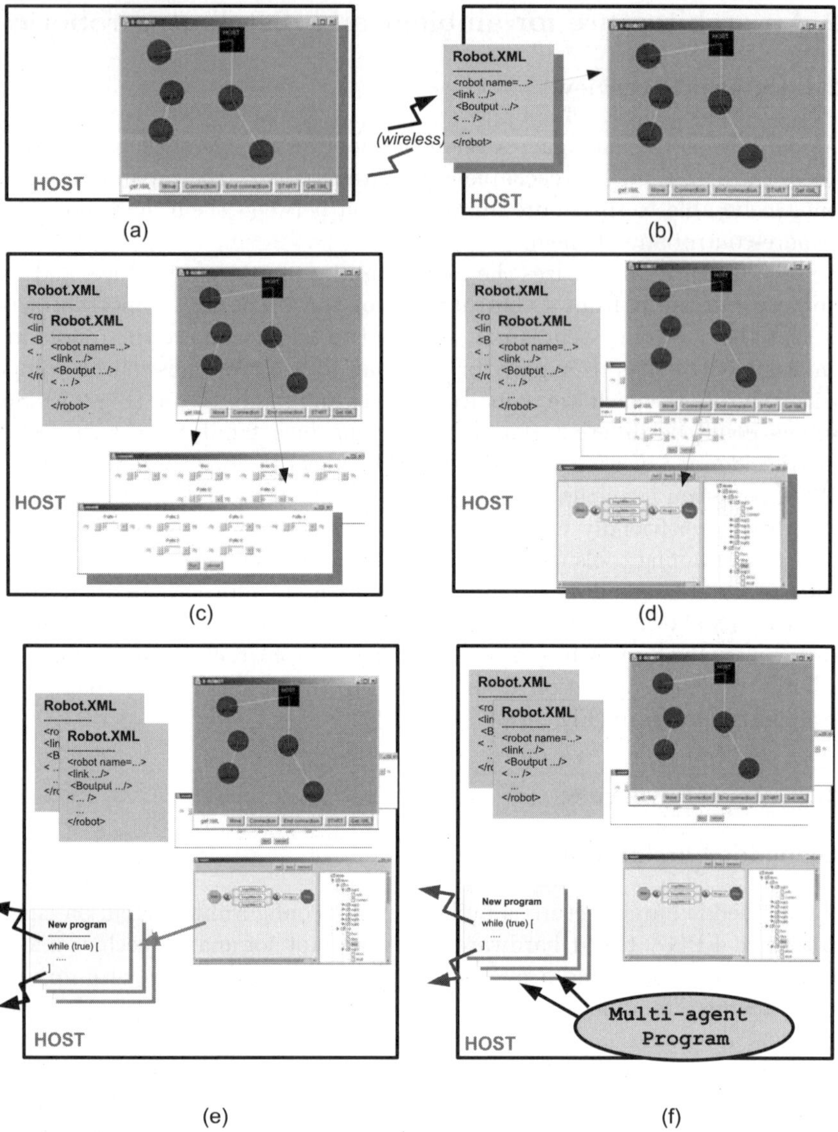

Fig. 2. Ambient behavior. (a) The host inquires for robots in its area. (b) The XML description of each robot is requested by the host. all the robots send their own XML files. With these informations the host (c) will be able to generate direct command on one robot or programs over a set of robots. (d) While the XML description contains elements of the local language of the robots, programming tools can be build. (e) New programs in local language are upload in the robots. (f) All these operations can be centralized in a global multi-agent program.

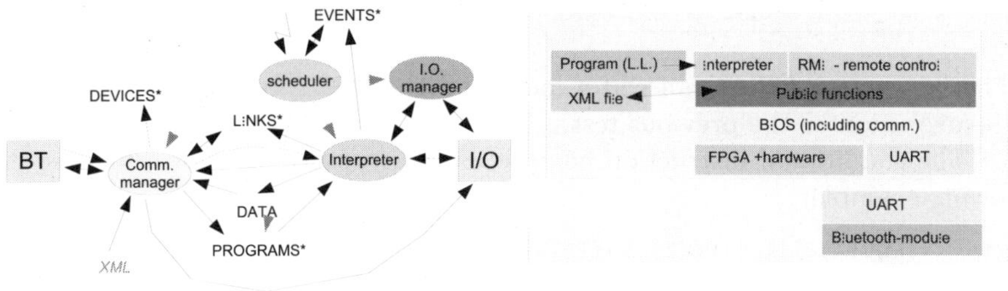

Fig. 3. BIOS. Data flow view (left) and layered view of the embedded software.

2.3 The highest layer: an interpreter

General structure of a program

A program is composed with a list of event, a list of variables and a list of instructions. Some of the instructions are specific to the agent and coincide with the embedded XML description. The control structures are the same.

Example

```
{K=0}
<SUP(can6,10):On2(1);>
<SUP(can3,20):On2(2);>
SET(K,-50);
EVT();
*[TRUE](    WAIT(50); INC(K);
            [EQU(K,50)]((SET(K,-50)!(PwmA(1,K);));
       );
```

In this example, two events are considered. `can6` and `can3` are internal registers implemented in the language. In this case they are the values of analog inputs, so these events aim at controlling the threshold of two sensors. The instruction `EVT()` starts the event scheduler.

Overview of the language

The syntax for instructions is the same as in many languages. All parameters and return values are integers. All basic operators are provided: arithmetic (`ADD`, `SUB`, `INC`, `DEC`, `MUL`, `DIV`); relational (`INF`, `SUP`, `EQU`, `NEQ`, `LEQ`, `GEQ`); logic (`AND`, `OR`, `NOT`) and affectation `SET`.

Control structures

The following examples gives three usual situations:

`{X=255, K=0}`	var: X $\leftarrow$ 255 and K $\leftarrow$ 0
`100*( PwmA(1,X); WAIT(10); DEC(X); );`	repeat 100 times
`*[LEQ(K,127)]( SET(K,Analog(1)); );`	while K(analog input)$\leq$127 do ()
`*[TRUE]( On(1); Off(1);)`;	while (true) do ()

The instruction BREAK can end a loop, and the CONT (for *continue*) redirect the execution to the previous test.

The conditional instruction has the usual structure, as shown in the following example:

```
[EQU(X,0)](  (On(1);WAIT(10);)      // executed if X==0
            !(Off(1);WAIT(20);)     // else ...
           );
```

2.4 Instruction for communication

The major originality of this local language is to incorporate special native instructions for communication. They can be divided in four groups:

1. Control of local unit:
   RSTBT(); → resets the *bluetooth* module;
   ST("name/message"); → changes the user-friendly name of the *bluetooth* module; since this string can be read without opening a link between agents, we use it as a mail-box. An example of name should be "a1/2,6,7,9" meaning (i) that the name of this robot is "a1", and (ii) that from "a1", the four robots "a2", "a6", "a7" and "a9" are reachable. So the host can initiate an *ad-hoc* graph without any connexion at the application level.
2. Link management:
   SET(Id,OPEN(address)); → opens a data link between the local module and the given address. The link number is stored in the variable Id.
   WCX(); → waits for a connection, this agent has nothing else to do.
3. Signals management:
   SYN(K); → waits for the signal K, incoming from any agent;
   SIG(Id,K); → sends the signal K towards the opened data–link Id.
4. Remote invocations: assume that the set of IO instructions of the agent is

   $$\mathcal{E} = \{\mathtt{INS}_k\}_{k\in[0,N]}$$

   then, this agent can apply all these instructions to a remote identical agent, after a successful connection. So the instructions of the set

   $$(r \circ \mathcal{E}) = \{\mathtt{rINS}_k\}_{k\in[0,N]}$$

   are implicitly integrate part of the language. They take as first parameter the identification of the target. The figure 4 shows an example of distributed action where a robot controlled by the host (planification level) is ordered to control three other robots. We can note the local instruction PwmA(x,y) and the remote invocation rPwmA(Id,x,y), where Id identifies an opened data–link.

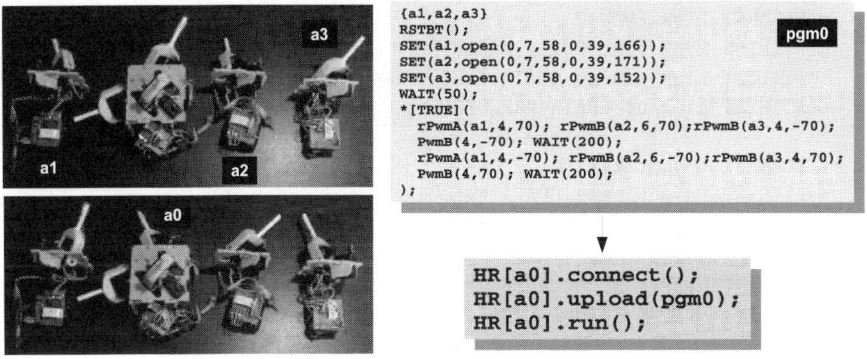

Fig. 4. Communication between agents. The host uploads a program in the agent a0 and launches it. Then, a0 take control of a1,a2 and a3.

2.5 Inner XML card

A DTD for mechatronical systems

The DTD is a set of grammar rules specifying the document generic logical structure. Thus in the DTD we enumerate all generic operations that concern the robots. Because we want easy transfers of this technology, we impose that (i) the number of kind of operation must be limited, and; (ii) the description can induce an object vision of the robot (a robot can be *composed* of subsystems themselves decomposable ...).

According to the first constraint, we propose only four kinds of functions:

$$\begin{cases} \text{binary outputs} \\ \text{outputs in a range} \in [min, \cdots, max] \\ \text{inputs} \\ \text{general procedures and functions} \end{cases}$$

A response for the second constraint is to add a concept of entity in the DTD and provide links between entities. So we can associate an operation with one of its subsystem. Because robots are often composed of many identical subsystems an entity can be link to n sub-systems (ex: hexapode, composed of 6 legs) Here is an extract of this DTD.

```
<?xml version='1.0' encoding='utf-8'?>
<!-- interfaceRobot.dtd -->
<!ELEMENT Robot (Link*,BOutput*,Output*,Input*,
                              Function*, Load?)>
<!ATTLIST Robot name CDATA #REQUIRED>
<!ATTLIST Robot BT CDATA #IMPLIED>

<!-- Logical link between objects -->
```

```
<!ELEMENT Link EMPTY>
<!ATTLIST Link name CDATA #REQUIRED>
<!ATTLIST Link parent CDATA #REQUIRED>
<!ATTLIST Link nb CDATA #REQUIRED>

<!ELEMENT Output EMPTY>
<!ATTLIST Output name CDATA #REQUIRED>
<!ATTLIST Output min CDATA #REQUIRED>
<!ATTLIST Output max CDATA #REQUIRED>
<!ATTLIST Output prot CDATA #REQUIRED>
<!ATTLIST Output parent CDATA #IMPLIED>

  ...
```

The `name` attribute concerns all the operations and entities of the robot. The `Link` element relies one system to n subsystems. When an operation addresses a subsystem, it fills the `parent` field.

The `prot` field appears for each operation element. Since we want to provide remote control tools and allow execution of distributed algorithms in a team of robots, we must access to all the internal basic functions of each robot. So this required field is necessary for defining the protocol of communication between the host and the robot.

XML example

The following code is an example of XML description using the previous DTD:

```
<?xml version="1.0" encoding="UTF-8"?>
<!DOCTYPE Robot SYSTEM 'interfaceRobot.dtd'>
<Robot name="H2" BT="07003F273A">
  <Link name="leg" parent="H2" nb="6"/>
  <Output name="PwmA" parent="leg" min="-115" max="115" prot="A"/>
  <Output name="PwmB" parent="leg" min="-115" max="115" prot="B"/>
  ...
```

The control system of the team of robots requests for all the XML files of the detected robots, and parse them in order to generate instructions for them. By reading this file, we guess that the robot H2 is composed of 6 legs, each of them is powered by two PWM commands. The next section shows the use of these informations.

3 Implementation in `maam` robot

3.1 Overview `maam` project

The `maam` [3] project is a self-reconfigurable robotic architecture where each module is autonomous for energy and CPU. The basic unit (called atom) is composed of six legs which are directed towards the six orthogonal directions of space. They allow the atom move itself and/or couple to another one. The first walking prototypes of atom appears on the figure 5.

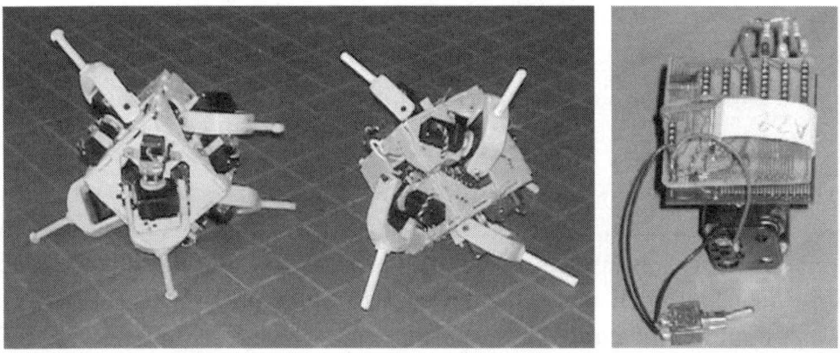

Fig. 5. The first prototypes of `maam` robot walking (right). These prototype embeds all the electronic and software functions described in this paper. They do not include the pincers.

The twelve PWM signals, and the command for the analog converter (driven in pipeline mode) are built in the FPGA. Some internal signals of the servo are processed in order to identify the legs in contact with the ground. Each leg is fitted with an IR transmitter/receiver for perception and docking. All the features described in this paper (XML inside, local language, wireless properties) are implemented and tested.

Because this robot do not take any inspiration from human or animal world a realistic simulator is written to study their behavior. The virtual robots have the same properties than the real robots and, moreover, are also driven by identical interpreted programs. The figure 6 reports a simulation were eight robots are searching for an attractor.

4 Conclusion

In this paper we presented an architecture for distributed robotic, completely abstracted by a language approach. Moreover, we proposed that every agent is fitted with an inside XML card, and that tools for control/command (including the dedicated local language) can be guessed thanks to this description.

[3] Molecule = Atom | (Atom+Molecule)

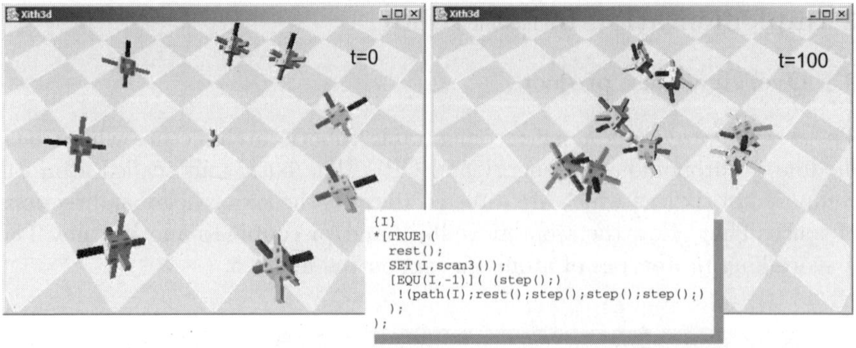

Fig. 6. Simulator. Like in the real context, the agents are independent and their own thread runs a program written in the interpreted local language.

The usual requirements in synchronization an remote invocation from one agent to another appear as a natural part in the language, since the middleware in charge of communication acts as a background task. The efficiency of this approach has been proved during its implementation in the `maam` robot which is a complex mechatronic system.

Acknowledgment

The `maam` project is supported by the Robea project of the CNRS. All references to people participating to this work can be found in [5].

References

1. Raja Chatila, "Control Architecture for Automous Mobile Robots",From Perception to Action Conference (PERAC'94), Lausanne , 1994, pp.254-265.
2. Medeiros, Adelardo A. D. "A survey of control architectures for autonomous mobile robots." J. Braz. Comp. Soc., Apr. 1998, vol.4, no.3. ISSN 0104-6500.
3. Dominik Henrich, Thomas Höniger, "Parallel Processing Approaches in Robotic", ISIE (1997)
4. Claude Guéganno and Dominique Duhaut , "A Hardware/Software Architecture for the Control of Self-Reconfigurable Robots", DARS 2004, Toulouse (France).
5. http://www.univ-ubs.fr/valoria/duhaut/maam

A Particle Swarm-based Mobile Sensor Network for Odor Source Localization in a Dynamic Environment

Wisnu Jatmiko*, Kosuke Sekiyama** and Toshio Fukuda*

*Department of Micro-Nano Systems Engineering, Nagoya University, 464-8603, Japan.
**Department of Human and Artificial Intelligence Systems, Fukui University 3-9-1 Bunkyo Fukui, 910-850, Japan.

Abstract. This paper addresses the problem of odor source localization in a dynamic environment, which means the odor distribution is changing over time. Modification Particle Swarm Optimization is a well-known algorithm, which can continuously track a changing optimum over time. PSO can be improved or adapted by incorporating the change detection and responding mechanisms for solving dynamic problems. Charge PSO, which is another extension of the PSO has also been applied to solve dynamic problem. Odor source localization is an interesting application in dynamic problem. We will adopt two types of PSO modification concepts to develop a new algorithm in order to control autonomous vehicles. Before applying the algorithm in a real implementation, some important hardware parameters must be considered. Firstly, to reduce the possibility of robots leaving the search space it is needed to limit the value of vector velocity. The value of vector velocity can be clamped to the range [$-V_{max}$, V_{max}]; in our case for the MK-01 Robot, the maximum velocity is 0.05 m/s. Secondly, in PSO algorithm standard there is no collision avoidance mechanism. To avoid the collision among robot we add some collision avoidance functions. Finally, we also add some sensor noise, delay and threshold value to model the sensor response. Then we develop odor localization algorithm, and simulations to show that the new approach can solve such a kind of dynamic environment problem.

Keywords. Particle Swarm Optimization, Odor Source Localization, Dynamic Environment

1. Introduction

The amount of research in the field of robotics application for odor-sensing technology has grown substantially. This work can be broadly categorized into two groups namely artificial odor discrimination system [1,2], and odor source localization by autonomous mobile sensing systems [3]. The artificial odor discrimination system has been developed for automated detection and classification of aromas, vapors and gases. The second prime area of robotics applications for odor-sensing technology is odor source localization. Odor source localization can be used for various attractive applications, including the search for detection of toxic gas leak and the fire origin at its initial stage, etc. This paper will address the second area of applications.

There have been several reported implementations of odor source localization by autonomous mobile sensing system. Most work on chemical sensing with mobile robots assume an experimental setup that minimizes the influence of turbulent transport by either minimizing the source-to-sensor distance in trail following [4,5] or by assuming a strong unidirectional air stream in the environment [6-9], including our previous work [10]. However, not much attention has been paid to the natural environment problem.

There has been no real implementation on a mobile robot that works in the natural environment to the best of our knowledge. The main problem in implementing odor source localization using a gas sensor in real world environments is that the distribution of the odorant molecules is usually dominated by turbulence rather than diffusion, the latter of which is known to be a considerably slower transport mechanism for gases in general. The other problem is the influence of unstable wind. When odor distribution is very complex and the wind direction is not stable, the robot will be haphazard and desultory [3].

This paper focuses on our new approach that exploits particle swarm optimization with multiple robots to solve odor source localization in natural environment where the odor distribution will change over time. Particle Swarm Optimization (PSO) simulates behaviors of bird flocking. Suppose the following scenario: a group of birds is randomly searching food in an area. There is only one piece of food in the area being searched. Not all of the birds know where the food is. However, they know how far the food in each iteration. So what is the best strategy to find the food? The effective approach is to follow the bird, which is nearest to the food. PSO learned from the scenario and applied it to solve the optimization problems [9]. However, the main problem with standard PSO used for dynamic optimization problems appears to be that PSO eventually will converge to an optimum and thereby looses the diversity necessary for efficiently exploring the search space and consequently the ability to adapt to a change in the environment when such a change occurs.

Two ways of improving PSO to solve this problem will be developed. Firstly, PSO is run in standard fashion, but when change in the environment has been detected, explicit actions are taken to increase diversity and thus to facilitate the shift to the new optimum. Therefore, PSO can be improved or adapted by incorporating change detecting and responding mechanisms for solving dynamic problems [12,13]. Secondly, multiple populations are used, some to track known local optima, some to search for a new optima. Two types of robot swarms, neutral and charged robots, will be used for solving the dynamic problem [14]. Odor source localization is an interesting dynamic problem application. We will adopt these two types of PSO modification concepts as described above to develop a new algorithm to control autonomous vehicles. These two types of PSO then will be compared for solving odor source localization in a dynamic environment.

2. Motivation

From early 1990 we developed real single mobile robot for solving odor source localization in natural environment. And also we developed simulation tool to implement several of sophisticated algorithm which we adopt from biological inspiration. In fact our system only can solve odor source localization with many simplicity parameters, like, stable wind and indoor environment [3, 10].

In the case of using mobile robots and multiple sensory modalities (e.g., odometry, anemometry, olfaction), we should carefully consider the feasibility of the hardware [15]. PSO, which incorporates change detecting and responding mechanisms, can be implemented with a simple algorithm in actual hardware.

Charged PSO, which employs two types of robots, neutral and charged robots, can also be implemented with a simple algorithm. With multiple populations, we can maintain the diversity of swarm particles. Applying a notion of electric potential field, charged swarm particle is introduced to make a balance of diversity. The potential field method is widely used in autonomous mobile robot path planning due to its elegant mathematical analysis and simplicity. The goal of this model is to have a number of sub-populations explore the best local optima. For this purpose, a part of the population is split off when a local optimum is discovered, and remains close to the optimum for further exploration. The remainder of the population continues to search for new local optima, and the process is repeated until better solutions are found. While the neutral swarm particles continue to optimize, the surrounding charged swarm particles maintain enough diversity to cope with dynamic changes in location of the covered peaks.

Evaluation on solving odor source localization problem in dynamic environment requires hardware and software platforms [16]. During the initial design stages, software evaluation is preferred, since such tools allow competing strategies to be evaluated under identical conditions for various environmental scenario. This paper presents a simulation implementation that addresses the tradeoffs between computational efficiency and inclusion of realistic hardware parameters.

3. Particle Swarm Optimization Frame Work

Recently, evolutionary techniques such as PSO have been applied to dynamic problem [11-14]. PSO can be improved or adapted by incorporating change detecting and responding mechanisms [12,13] for solving dynamic problem. CPSO, which is another extension of PSO, has also been applied to solve dynamic problem [14]. We will adopt concepts from modification two type of PSO as described above to develop a new algorithm to control autonomous vehicles.

3.1 Particle Swarm and Robot Interactions

A more detailed interaction of robot with Particle Swarm Optimization algorithm will be described as follows. A population of robots is initialized with certain positions and velocities and a function (plume distribution) is evaluated, using the robot's positional coordinates as input values. Positions and velocities are adjusted and the function evaluated with the new coordinates at each time step. When a robot discovers a pattern that is better than any it has found previously, it stores the coordinates in a vector P_i. The difference between P_i (the best point found by i so far) and the individual's current position is stochastically added to the current velocity, causing the trajectory to oscillate around that point. The stochastically weighted differences among the population's best position P_g and the individual's current position are also added to its velocity, adjusting it for the next time step. These adjustments to the robot's movement through the space cause it to search around the two best positions.

The values of element in P_g (concentration gas and position of robot) are determined by comparing the best performances of all the members of population, defined by indexes of other population members and assigning the best performer's index to the variable g. Thus, P_g represents the best position found by all member of population. Ad-hoc wireless network and Global Position System are assumed to be equipped among all robots. Via the ad-hoc network, each robot can collect the gas concentration data and choose the best one. Then the position of the robot can be determined by GPS system.

The PSO model is described as following:

$$V_i^{n+1} = \chi\left\lfloor V_i^n + c_1.rand().(p_i^n - x_i^n) + c_2.Rand().(p_g^n - x_i^n)\right\rfloor \quad (1)$$

$$x_i^{n+1} = x_i^n + V_i^{n+1} \quad (2)$$

After finding the two best values, the particle updates its velocity and positions with eq. (1) and (2). Let X_i and V_i denote position and velocity vector of the i-th particle at the iteration time n (n=1,2...). Also p_i and p_g are defined as the local best and the global best respectively as stated above. Rand () and rand () are the random functions returning a value between (0,1). Coefficient χ is constriction factor, which is supposed to take less than 1. Also coefficient c_1 and c_2 are learning parameters, which are supposed to be $c_1 = c_2 = 2$. All of the parameters are referred to [11, 12, 13], which was the best of PSO parameters in common optimization problem.

In the dynamic environment the standard PSO cannot solve the problem [11]. In order to solve the dynamic problem, PSO has to be improved or adapted by incorporating change detecting and responding mechanisms [12, 13]. Detection function is used for monitoring the global best information. If it has not changed for certain number of iterations, there supposed to be a possible optimum change. After the detection of environment changes, there must be an effective strategy to respond to a wide variety of changes. However, if the whole population of robot has already converged to a small area, it might not be easy to jump out to follow the changes. Therefore, we investigate the spreading response when a change is detected. All robots will spread at a certain step to jump out to follow the changes, for simplicity reason [12, 13].

3.2 Multi Swarm Robot Interaction

With the charged swarm robots we add repulsion function to make balancing diversity (like potential field idea). This keeps robots from gathering at a small area. The charged swarm robot will enable to explore different regions of the search space by different swarm. Charged swarm robots are adopted from the concept of Coulomb's law.

Figure 1 shows the repulsion function for charged swarm robots. Additional avoidance is only between pairs of robots that have non-zero charge Q (charged robot), and the radius of the avoidance was adopted from Coulomb's law, like in the shell $r_{core} \le r \le r_{perc.}$ At separations less than the core radius r_{core}, the repulsion is fixed at the value at the core radius, and there is no avoidance for separations beyond the perception limit of each robot r_{perc}.

$$\vec{a}_i = \Sigma\, a_{i,p} \qquad \rightarrow \text{Summation of repulsive force}$$

$$a_{i,p} = \begin{cases} \dfrac{Q_i \cdot Q_p}{\left|x_i - x_p\right|^3}(x_i - x_p) \ldots r_{core} < \left|x_i - x_p\right| < r_{perc} & \text{"Region} - 2\text{"} \\ \dfrac{Q_i \cdot Q_p (x_i - x_p)}{r_{core}^{\,2}\left|x_i - x_p\right|} \ldots\ldots\ldots\ \ldots\ldots\ \left|x_i - x_p\right| < r_{core} & \text{"Region} - 1\text{"} \\ 0 \ldots\ldots\ldots\ \ldots\ldots\ldots\ \ldots\ldots\ldots\ \ldots\ldots\ldots\ \left|x_i - x_p\right| > r_{perc} & \text{"Region} - 3\text{"} \end{cases} \tag{3}$$

Two types of robot swarm can be defined as neutral and charged robots. The neutral swarm robots have no charged function and identical with the standard PSO, as described in eq. (1) and (2). However, in charged swarm robot, there is an additional term to facilitate collision avoidance.

The charge swarm robot is described in eq. (4) and (5).

$$V_i^{n+1} = \chi\left[V_i^n + c_1.rand().(p_i^n - x_i^n) + c_2.Rand().(p_g^n - x_i^n)\right] + \vec{a}_i \tag{4}$$

$$x_i^{n+1} = x_i^n + V_i^{n+1} \tag{5}$$

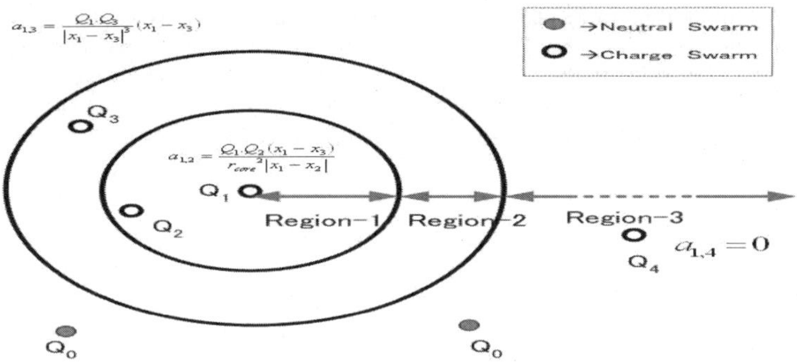

Figure 1 Charged Swarm Robot Interaction

3.3 Algorithm Implementation

The problem of gas source localization in an enclosed 2D area can be decomposed into three subtasks: plume finding (coming into contact with the plume), plume traversal (following the plume to its source) and source declaration (determining the source is in the immediate vicinity).

We used a random search until one robot getting into contact with the plume. After getting into contact with the plume, the second task is plume traversal. Particle Swarm concept will be applied for following the cues determined from the sensed gas distribution toward to the source. The last task is source declaration, determining certainty that the gas source has been found. We also used Particle Swarm Optimization convergence parameter, which is to find global maximum, which value is known.

4. Simulation Experiment

4.1 Environment

The Gaussian plume model was adopted from J. O. Hinze [17] and Ishida [18]. The Gaussian gas distribution is expressed by:

$$C(x,y) = \frac{q}{2\pi.K.d_s} \exp\left[-\frac{U}{2K}(d_s - \Delta x) \right] \tag{6}$$

where,

$$d_s = \sqrt{(x_s - x)^2 + (y_s - y)^2} \tag{7}$$

$$\Delta x = (x_s - x)\cos\theta + (y_s - y)\sin\theta \tag{8}$$

C is concentration of plume (ppm), q is emitted rate of the gas (mL/s), U is wind speed (m/s), K is turbulent diffusion coefficient in m^2/s, θ is the angle from the x-axis to the upwind direction.

4.2 Robot Behavior

When applying the algorithm in a real implementation, some important parameters should be considered. Firstly, in order to reduce the possibility of robots leaving the search space it is needed to limit the value of velocity vector. The value of velocity vector can be restricted to the range $[-V_{max}, V_{max}]$; in our case for the MK-01 Robot, the maximum velocity is 0.05 m/s. Secondly, in PSO algorithm standard there is no collision avoidance mechanism. To avoid the collision among robots we add some collision avoidance functions. Finally, we also add some sensor noise and threshold value to model the sensor response.

$$S(x,y) = C(x,y) + e(x,y) \tag{9}$$

$$S(x,y) = \begin{cases} S(x,y) & if \ldots C(x,y) > \tau \\ 0 & otherwise \end{cases}$$

Where S(x,y) is the sensor's response, C(x,y) is gas concentration, e(x,y) is the sensor noise with e(x,y) << C(x,y) and τ is 1 ppm.

4.3 Typical Approach

Firstly we apply the PSO approach in static environment (the plume is very stable), to show the robot interaction. The parameter settings for PSO algorithm were fixed at standard value [12,13] and we will use ten robots for all the experiments. As shown in Fig. 2, three sub-tasks of algorithm will be applied to solve the static environment and we can see the approach can easily find the odor source.

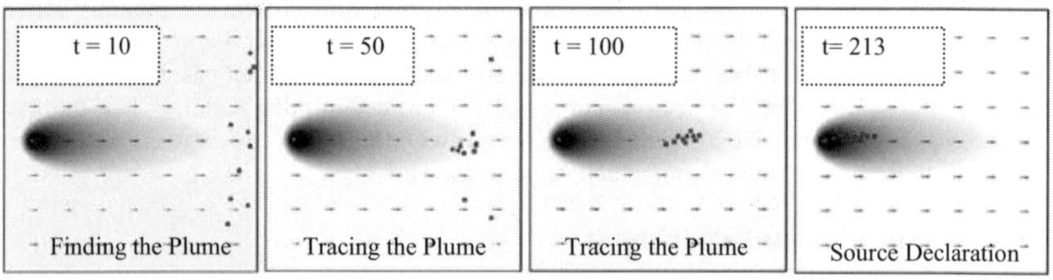

Figure 2 Visualization of proposed approach with three sub-tasks: Plume Finding, Plume Traversal and Source Declaration

4.4 Experiment Result in Dynamic Environment

In reality many real-world problems are dynamic, like odor localization in natural situation is also dynamic. The next experiment we apply the PSO approach in dynamic environment (the plume is not stable). The plume changes randomly according to the wind speed and wind direction.

In our simulation the wind speed changed randomly from 0.5 m/s to 1 m/s and the wind direction changed randomly from 160^0 to 200^0. The timer for make random changing is from 20 s until 50 s. Two types of modified PSO are used to solve the problem. Firstly, PSO incorporating change detecting and responding mechanisms and secondly Charge PSO.

In dynamic environment the standard PSO can not solve the problem. As we can show in figure 3, the robot was trapped in local maximum area. Experimental result of detect and respond PSO is shown in Fig. 4. Detection function use for monitoring the global best information, if it has not changed for 20 iterations, there is a possible optimum change and the value of global best will be restarted at the initial value (global best=0). After the detection of environment changes, we investigated the spreading response when a change is detected. All robots will spread at 5 steps to jump out to follow the changes.

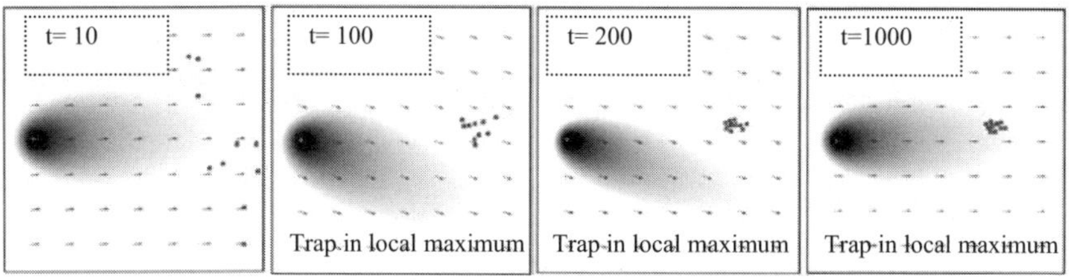

Figure 3 The propose approach with standard PSO can not follow the changing of environment, trap in local maximum.

The effect of restarting the global best and spreading implies loss of information gathered during the search so far. As an alternative adaptation, Charge PSO approach is investigated. Experiment result of Charge PSO is shown in Fig. 5. The parameters for charge robot used in the experiment are charge value Q=1, r_{core} =1 meter and r_{perc}=2 meter.

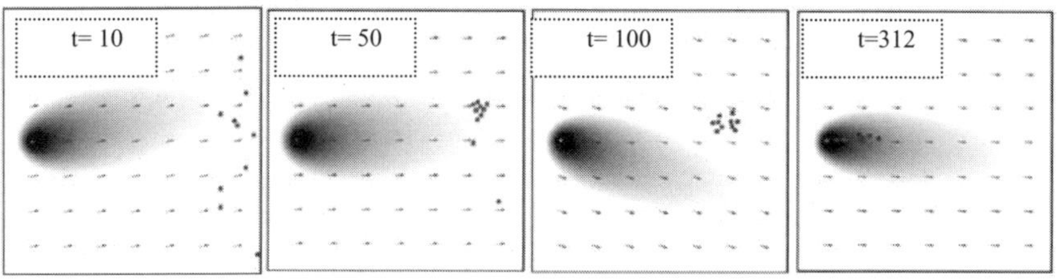

Figure 4 The proposed approach with Detect and respond PSO can follow the changing of environment.

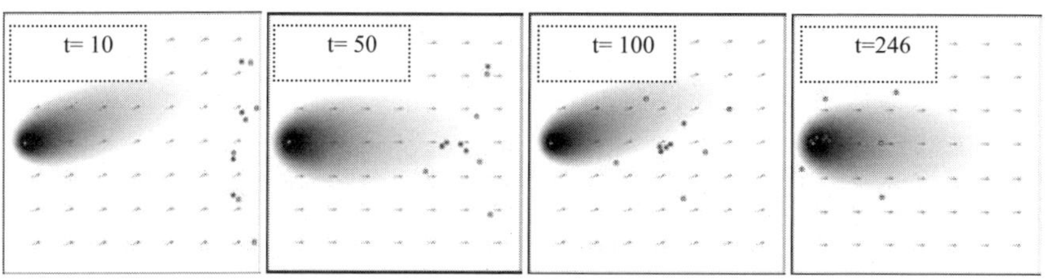

Figure.5 The proposed approach with Charged PSO can follow the changing of environment.

4.5 Analysis

It is important to analyze the results from several repeated runs by statistical methods, in order to obtain empirical evidence of the capabilities of proposed method. We analyze the efficiency of the proposed method which expressed as the number of iterations to find the solution. Then with MEAN statistical analysis, we measure the performance.

Figure 6 shows the compared results among PSO algorithms modification for solving odor source localization in dynamic environments. The PSO standard can not solve the problem until the last iterations. The Detect and Respond PSO can solve the odor source lo-

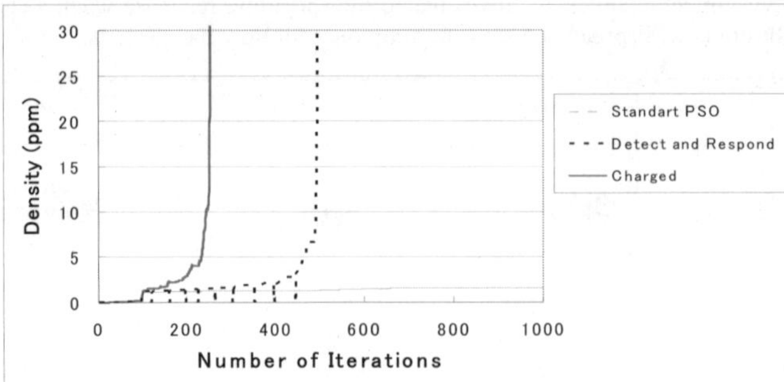

Figure 6 Time development of global best coping with dynamical change of environment with various PSO algorithms. (Taking by 25 times)

calization problems however the drawback is with the arbitrary nature of the detection and response algorithms. Particle Swarm with charge need no further adaptation to cope with dynamic scenario due to the extended swarm shape. The next experiment we just concern with the Charged PSO algorithm. Figure 7 shows the results the time development of global best coping with dynamical change of environment when delay and noise sensor was employed.

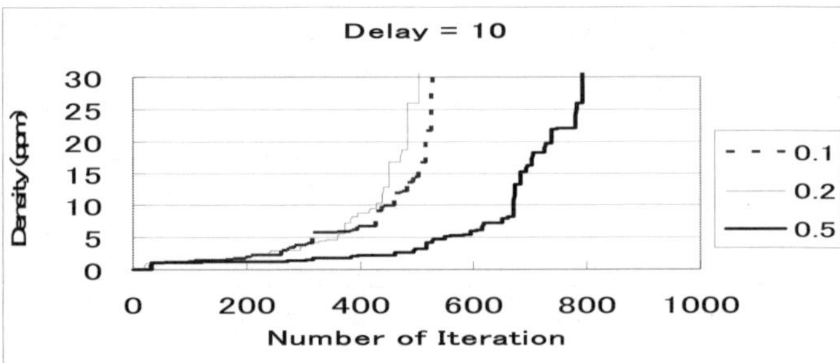

(a)

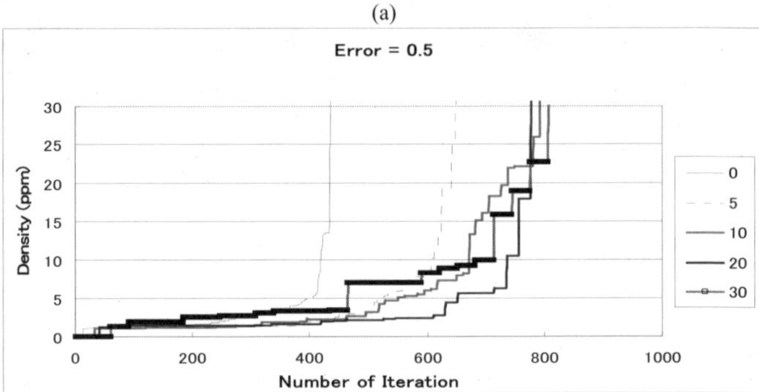

(b)

Figure 7 Time development of global best coping with dynamical change of environment used Charged PSO algorithm with employed uncertain sensor parameters (a). with delay 10 s and various error, i.e., 0.1, 0.2 and 0.5 (b). with error 0.5 and various time delay, i.e., 0 s, 5 s, 10 s, 20 s and 30 s. (Taking by 25 Times)

We also used another advanced turbulence model by Farrell et al [16], chosen because of its efficiency, realism (i.e., its instantaneous and time-averaged results much measurement of actual plume), and multi-scale properties – including chemical diffusion and advective transportation. Figure 8 shows our approach can solve the advanced turbulence model.

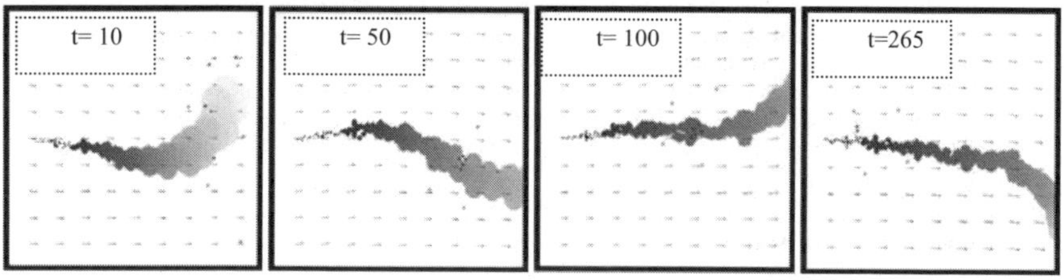

Figure 8 The proposed approach with Charged PSO can follow the changing of environment in advance turbulence environment.

5. Conclusions and Future Work

In this paper, we have presented two types of Particle Swarm Optimization modification approaches to control autonomous vehicle robots to search for odor source in dynamic environment. The Detect and Respond PSO can solve the odor source localization problems however the drawback is with the arbitrary nature of the detection and response algorithms. Particle Swarm with charge need no further adaptation to cope with dynamic scenario due to the extended swarm shape.

Although odor source localization can be used for various attractive and promising applications, so far there have been few applications on odor source localization by autonomous mobile sensing system in real world environment. The main problem to implement odor source localization with using gas sensor in real world environments is that the distribution of odorant molecules is dominated usually by turbulence rather than diffusion, which is known to be considerably slower transport mechanism for gases in general. The other problem is unstable wind in real world environment. Beside that, our understanding of the solving odor source localization, particularly in dynamic environment, is still in its infancy. In real natural environment the robot will find variety of situation which related multi study from biology, physic-chemistry, engineering and robotic. Unresolved problem still find in implementation phase. Most of those could be grouped into one of the following categories:

1. Environment:
   Environment with various obstacles, as a result, the environment become more realistic and complicated. And also the changing of wind direction, yet, the changes in wind's direction are very are very limited (only 40 degrees) in comparison to the changes possible in natural environment (up to 180 degree).
2. Algorithm optimization:
   The common problem using PSO is a tuning parameter for find the optimal solution. Most of the researcher use, a cross validation or try and error tuning parameter. Further algorithm development in simulation will include online learning (parameter) which system can learn from environment. As our experience used EA in another problem of odor-sensing implementation, that is artificial odor discrimination system [1, 2].
   Considering an application to robots, a decision making architecture, which synthesizes interactions for odor source detection and cooperative mobile behavior under constraints (i.e. error model for GPS sensors), has to be present. Another important factor in PSO is neighborhood topology. In our approach we used fully connected neighborhood topology. In the paper by Kennedy [11], other topologies are described as: 1) Circle Topology and 2) Wheel Topology. We also try to analyze the feasibility conjectures referred to above, in future work.
3. Real Hardware Implementation:
   An important near-term focus will be on porting the simulation to actual robots (MK-01) in a laboratory experiment. Multiple autonomous mobile robots developed by Fuji Heavy Industries Ltd.

will use for actual robot experiment. This robot can move autonomously, that has 16 middle range infrared sensors, eight close range infra red sensor, two actuators, a microcontroller. Additionally, a robot can communicate each other by wireless LAN. Our robot used TGS-822 gas sensor for alcohol and volatile vapor detection from Figaro Inc. The sensing element of TGS-822 gas sensors is a tin dioxide (SnO2) semiconductor that has low conductivity in clean air. In the presence of a detectable gas, the sensor's conductivity increases depending on the gas concentration in the air. A simple electrical circuit can convert the change in conductivity to an output signal which corresponds to the gas concentration. The TGS-822 has high sensitivity to the vapors of organic solvents as well as other volatile vapors. It also has sensitivity to a variety of combustible gases such as carbon monoxide, making it a good general purpose sensor. Via the ad-hoc wireless LAN, each robot can collect the gas concentration value and choose the best one. Then the position of the robot can be determined covering camera.

Acknowledgments

The authors are grateful to Prof. J. A. Farrell from University of California, Riverside, U.S.A., for his support advance turbulence environment source-code.

References

[1] B. Kusumoputro, H. Budiarto and W. Jatmiko," Fuzzy-Neural LVQ and Its Comparison with Fuzzy Algorithm LVQ in artificial odor discrimination system ," ISA Trans. Sci. Eng. Meas. Autom, Elsevier, vol. 31, pp.395-407, October 2002.

[2] W. Jatmiko, T. Fukuda, F. Arai and B. Kusumoputro, "Artificial Odor Discrimination System Using Multiple Quartz Resonator Sensor and Various Neural Networks for Recognizing Fragrance Mixtures, IEEE Sensors Journal., vol. 6. no.1, pp.223-233, Feb. 2006.

[3] W. Jatmiko, T. Fukuda, T. Matsuno, F. Arai and B. Kusumoputro," Robotic Applications for Odor-Sensing Technology: Progress and Challenge, WSEAS Transaction on System、Issue 7, Volume 4, July 2005

[4] M. Wandel, A. Lilienthal, T. Duckett, U. Weimar, and A. Zell. Gas distribution in unventilated indoor environments inspected by a mobile robot. In Proceedings of the IEEE International Conference on Advanced Robotics (ICAR'03), 2003.

[5] X. Cui, C. T. Hardin, R. K. Ragade, and A. S. Elmaghraby. A swarm-based fuzzy logic control mobile sensor network for hazardous contaminants localization. In Proceedings of the IEEE International Conference on Mobile Ad-hoc and Sensor Systems (MASS'04), 2004.

[6] Ishida, H. Nakayama, G. Nakamoto and T. Moriizumi, T. Controlling a gas/odor plume-tracking robot based on transient responses of gas sensors", IEEE Sensors Journal, Vol. 5. No.3. June 2005.

[7] Adam T. Hayes, A. Martinoli and R. M. Goodman, "Distributed Odor Source Localization," IEEE Sensors Journal, Vol. 2. No.3. June 2002.

[8] D. Zarzhitsky, D. Spears, and W. Spears. Distributed Robotics Approach to Chemical Plume Tracing. Proceedings of the IEEE/RSJ International Conference on Intelligent Robots and Systems (IROS'05), 2005.

[9] R. A. Russell and A. H. Purnamadjaja. Odor and airflow: Complementary senses for a humanoid robot. In Proceedings of the 2002 IEEE International Conference on Robotics and Automation, 2002.

[10] W. Jatmiko, B. Kusumoputro, and Yuniarto, "Improving the Artificial Odor and Gas Source Localization System Using the Semiconductor Gas Sensor Based on RF Communication", Proc. of IEEE APCASS, October 2002.

[11] Russell C. Eberhart, and James Kennedy, Swarm Intelligence, The Morgan Kaufmann Series in Artificial Intelligence,2001.

[12] Eberhart, R. C. and Shi, Y. "Tracking and optimizing dynamic systems with particle swarms." Proceedings of the IEEE Congress on Evolutionary Computation (CEC 2001), Seoul, Korea. pp. 94-97, 2001.

[13] X. Hu, and R. Eberhart. "Adaptive particle swarm optimization: detection and response to dynamic systems". Proceedings of Congress on Evolutionary Computation, 2002. pp. 1666-1670. Hawaii, USA.

[14] T. Blackwell and J. Branke. "Multi-swarm optimization in dynamic environments." In G. R. Raidl, editor, Applications of Evolutionary Computing, volume 3005 of LNCS, pages 489-500. Springer, 2004.

[15] Asama, H. Arai, T. Fukuda and Hasegawa T, editors. 2002, Distributed Autonomous Robotics Systems (DARS 5) Springer-Verlag, Berlin.

[16] Jay A. Farrel et all , "Filament-based atmospheric dispersion model to achieve short time-scale structure of odor plumes," Environment Fluid Mechanics , vol. 2, pp. 143-169, 2002.

[17] J.O. Hinze, Turbulance, McGraw-Hill, New York, 1995..

[18] H. Ishida, T. Nakamoto and T. Mpriizumi,"Remote Sensing of Gas/Odor Source Localization and Concentartion Using Mobile System", Sensors and Actuators B 49 (1998) 52-57.

Cooperative Multi-robot Target Tracking

Boyoon Jung[1] and Gaurav S. Sukhatme[2]

[1] NavCom Technology, Inc., 20780 Madrona Avenue, Torrance, CA 90503, USA
bjung@navcomtech.com
[2] University of Southern California, Los Angeles, CA 90089, USA
gaurav@robotics.usc.edu

Summary. Target tracking performance can be improved by using multiple robot trackers, but this requires a coordinated motion strategy among the robots. We propose an algorithm based on treating the densities of robots and targets as properties of the environment in which they are embedded. By suitably manipulating these densities a control law for each robot is proposed. The proposed algorithm has been tested through intensive simulations and a real-robot experiment. First, two different versions of the approach were evaluated by studying the performance change as the communication range among robots varies. The results showed that our treatment of the coordination problem is effective and efficient. Second, the developed system was tested on two Segway RMP robots, and the behaviors of the robots in a cooperative tracking experiment provide evidence that the proposed method controls multiple robots appropriately according to the target distribution change.

Key words: mobile robot, multi-robot system, cooperative target tracking.

1 Introduction

Using a mobile robot as a tracking device is beneficial because (1) a mobile robot can cover a wide area over time, which means the number of sensors required for tracking can be kept small, and (2) a mobile robot can re-position itself in response to the movement of the targets for efficient tracking. In cases where the number of targets is much larger than the number of sensors available or when sensors cannot be deployed in advance at the correct locations, mobility is indispensable. Tracking performance can be improved by using multiple robots, and this requires a coordinated motion strategy among robots for *cooperative target tracking*.

The multiple target tracking problem using multiple mobile robots is defined as follows:

Input Estimated poses of M robots and estimated positions of n tracked targets (out of total N targets) in a bounded environment E ($M \ll N$)

Output Motion commands for M robots

Goal Maximize the number of tracked targets n over time T

$$Observation = \sum_{t=0}^{T} \frac{n(t)}{N} \times \frac{1}{T} \times 100 \qquad (1)$$

Restriction No prior knowledge of the number of robots or targets, and no target motion model.

The problem seems well-suited to a POMDP optimization framework, and an optimal solution is guaranteed by solving the POMDP problem. However, this approach is not directly applicable to a real world system. The most critical limitation is that the size of the state space ($|E|^{M+N}$) increases exponentially as the number of robots or targets increases. Since the evaluation time of the POMDP problem is exponential in the size of the state space, the problem becomes intractable quickly. Therefore, for scalability, a *distributed* solution is preferable to the centralized, optimal solution. Another limitation is that the optimal policy needs to be re-computed whenever the system configuration changes (examples include adding or removing robots at run time, or adding/removing targets at run time) which implies that the policy computation should be done in real-time. However, most optimization techniques require a significant amount of computation and memory, and they are not suitable to real-time application. Therefore, an *on-line* algorithm is preferable to the off-line, optimal solution.

We propose a *Region-based Approach* as an efficient coordination method, which distributes robots according to the target distribution. In our approach, each robot broadcasts its location and the locations of currently tracked targets. Based on this information and similar information gathered from other robots, each robot independently maintains an estimate of two density distributions - the robot density and the target density. A control law for each robot is generated by using these density estimates. Communication among robots is the key enabler for multi-robot coordination, so the effect of communication range was analyzed; we observed the performance change as the communication range varies. The simulation results show that the proposed algorithm is efficient and robust. The proposed method is also implemented and tested using real robots to validate its applicability to a real-world, resource-constrained system.

The paper is organized as follows. Section 2 summarizes the related work on this topic, and the *Region-based Approach* algorithm is described conceptually in Section 3. Section 4 reports the experimental results and analyzes the performance of the proposed algorithm. The current status and possible improvements are discussed in Section 5.

2 Related Work

Various distributed algorithms have been proposed for multi-robot coordination with applications to multi-target tracking. The ALLIANCE architecture [8] achieves target-assignment by the interaction of motivational behaviors. If a target was not tracked for a while, the robot which was supposed to track the target would give up and another robot in a better position would take up the target. In the BLE architecture [12], if a particular robot thinks it is best suited to track a specified target, it stops other robots from tracking the target by broadcasting inhibition signals over the network. The Murdoch architecture [1] showed that target-assignment problem can be solved using a principled publish/subscribe messaging model; the best capable robot is assigned to each tracking task using a one-round auction.

There have been other approaches that control robot position without explicit target assignment, especially when the ratio of the number of robots to the number of targets is close to 1.0. In [9, 10], the configuration of a team of mobile robots was actively controlled by minimizing the expected error in tracking target positions, and a decentralised system architecture maximizing local information gains was presented in [2]. A reactive motion planner was reported in [7] that maximizes the shortest distance that a target needs to move in order to escape an observer's visibility region.

The Pursuit-Evasion problem introduced in [13] is a formally simplified tracking problem. The goal is to find continuously-moving intruders using a single or multiple searchers with flashlights that emit a single ray of light. [13] presents upper and lower bounds of the number of necessary searchers in a given environment (a simple polygon) and four measures of shape complexity of the environment (the number of edges, the number of reflex vertices, the bushiness, and the size of a minimum guard set). [3] extend the problem to exploit a visible area instead of a single ray of light. Several bounds on the number of pursuers are defined and the complete algorithm for a single pursuer case is presented.

3 Region-based Approach

The *Region-based Approach* is based on the following fundamental assumption:

> *For two comparably sized regions, more robots should be deployed in the one with the higher number of targets.*

Instead of allocating targets to each robot, robots are allocated to each region based on the target distribution and robot distribution. The *robot density* and the *target density* are defined for each position in an environment, and a robot is attracted to (or repulsed from) the position based on those density estimates. For example, the less targets a region has, the less robots the region requires, and the more robots the current region has, the more robots in that region are free to move to other regions. Our approach assumes the following:

Global Localization All robots share a global coordinate system so that the positions of targets detected by different robots can be translated into a single coordinates.

Robust Tracker The cooperative tracking algorithm is decoupled from the low-level target tracker; such a single-robot tracker is described in [5].

Bounded Environment The size of an environment is bounded by the communication range among robots or memory constraints, not by the intrinsic limitation of our algorithm.

3.1 Relative Density Estimates as Attributes of Space

In order to compute robot and target density values at each position, models for robot position, target position, and region boundary are required. Based on the output of the localization algorithm, the position of a robot can be modeled by a delta function or a Gaussian function. When the localization algorithm returns an exact position ($\mathbf{x}_i$) as the best estimate, the robot position is modeled using a delta function (Eqn. 2). When the localization algorithm returns a center position (μ_i) as the best estimate and a covariance matrix (Σ_i) as uncertainty estimate, a bi-variate Gaussian model (Eqn. 3) is adequate. The robot distribution r over an environment is computed by summing the individual models, which are collected through communication among robots at run-time.

$$r = \sum_i \delta(\mathbf{x}_i) \tag{2}$$

$$r = \sum_i N(\mu_i, \Sigma_i) \tag{3}$$

In similar way, the target distribution can be computed. Based on the output format of an underlying target tracker, a delta function model or a bi-variate Gaussian model can be used. The target distribution t over an environment is computed as follows:

$$t = \sum_i \delta(\mathbf{x}_i) \tag{4}$$

$$t = \sum_i N(\mu_i, \Sigma_i) \tag{5}$$

To define the density estimates, a region boundary R of a unit space must be defined. We consider two possibilities: a binary model and a Gaussian model. The binary model (Eqn. 6) defines a region boundary with radius r, which is conceptually simple and computationally cheap. A Gaussian model (Eqn. 7) can be used to define a region boundary when a differentiable output is preferred. The Gaussian distribution is zero-centered and the boundary is determined by a covariance matrix Σ.

$$R(\mathbf{x}) = \begin{cases} 1.0 & \text{if } |\mathbf{x}| < r \\ 0.0 & \text{otherwise} \end{cases} \tag{6}$$

$$R(\mathbf{x}) = N(\mathbf{0}, \Sigma) \tag{7}$$

The final density distribution of robots (D_r) or targets (D_t) is computed using a convolution of the robot location and region extent:

$$D_r(x, y) = r \otimes R = \int_{-\infty}^{\infty} \int_{-\infty}^{\infty} r(\tau, \rho) R(x - \tau, y - \rho) d\tau d\rho \tag{8}$$

$$D_t(x, y) = t \otimes R = \int_{-\infty}^{\infty} \int_{-\infty}^{\infty} t(\tau, \rho) R(x - \tau, y - \rho) d\tau d\rho \tag{9}$$

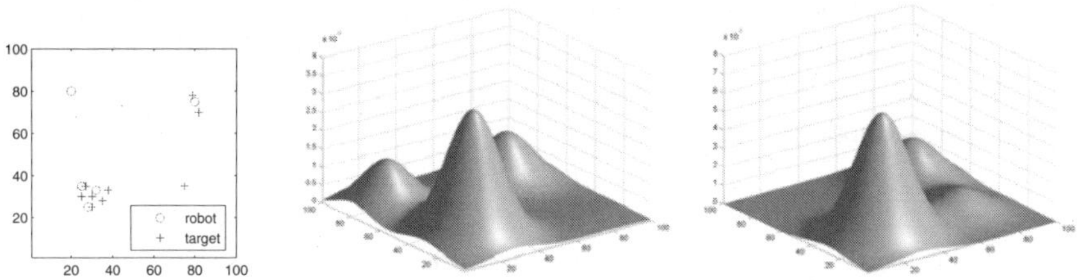

(a) object positions (b) robot density distribution (c) target density distribution

Fig. 1. Density distributions

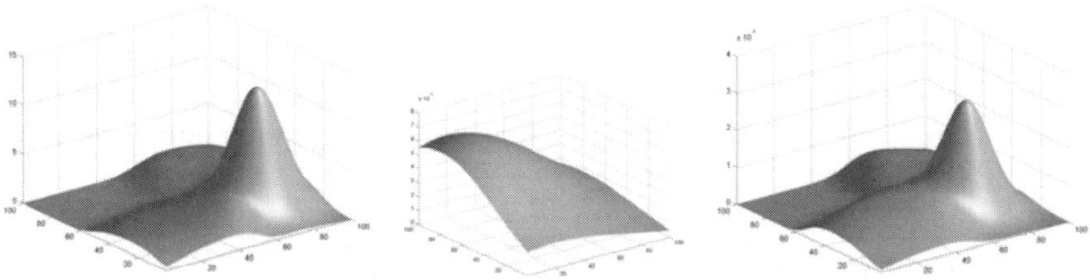

(a) urgency distribution (b) cost function (c) utility distribution

Fig. 2. Utility distribution

Figure 1 (b) and Figure 1 (c) show the final density distribution examples with Gaussian models (Eqn. 3, 5, and 7) when the positions of robots and targets are as shown in Figure 1 (a).

3.2 Urgency Distribution and Utility

Given the distribution of robots D_r and the distribution of targets D_t in a bounded environment, we define the urgency distribution u:

$$u(x, y) = \frac{D_t(x, y)}{D_r(x, y)} \tag{10}$$

Figure 2 (a) shows the urgency distribution calculated using Gaussian models. As shown in Figure 1 (a), there are three groups of targets: six targets around the coordinate $(30, 30)$, two targets around $(80, 80)$, and a single target at $(75, 35)$. The first two groups are being observed by robots, and those regions have relatively low urgency values as shown in Figure 2 (a). However, the last group is not being tracked by any robot, so the urgency value of the region is very high, which means the region 'requires' a robot to migrate towards it.

A cost function c_r for each robot can be combined to compute the final utility function for robot control instead of simply using the urgency distribution as an utility function. For example, the cost of motion can be factored in by multiplying a function which is inverse-proportional to the travel distance. Figure 2 (b) shows an example; the inverse cost function for the robot at $(20, 80)$ has a peak at the current position of the robot since the cost of traverse is zero, and it decreases as it moves further from the current position because the cost of traverse increases. The final utility distribution function is defined as:

$$U(x,y) = u(x,y) \times \frac{1}{c_r(x,y)} \tag{11}$$

It is worth noting that each robot maintains a utility distribution *independently*, and thus each robot would have a different utility distribution from others because of the cost function term c_r. Since the urgency distribution u is calculated using the position information of robots and targets, every robot would maintain the same u distribution when global communication is available. However, the different positions of robots cause different costs for a region, and eventually diverse behaviors for robots are generated.

The final utility distribution for the robot at the coordinate $(20, 80)$ is shown in Figure 2. Intuitively, the region at the coordinate $(75, 35)$ would attract the robot since it has the highest utility value.

3.3 Distributed Motion Strategy

Given the utility distribution, we define two motion strategies. If only local planning is desired, then one possible motor command is a gradient descent method on the utility function as follows:

$$\dot{\mathbf{x}} = -\nabla U \tag{12}$$

If global planning is preferred, then the peak position of the utility distribution can be a goal position:

$$\mathbf{x}' = \arg\max_{\mathbf{x}} U(\mathbf{x}) \tag{13}$$

Each robot plans its motion and executes it *independently* in a distributed manner, and there is no explicit negotiation between robots. However, by sharing the position information of robots and targets, these motion plans are coupled.

4 Experimental Results and Discussion

4.1 Effect of Communication Range

The most critical factor in the performance of the *Region-based Approach* is the communication range among robots. Since each robot estimates the robot and the target densities through communication and the control law is computed based on the estimates, the effectiveness of its motion depends critically on the accuracy of the estimates. Therefore, we studied how the performance of the proposed algorithm degrades as the communication range shrinks through intensive simulations.

The environment was a 50×50 meter sized empty space, and the grid size for the utility function representation was fixed to one meter. The number of robots and targets were fixed to three and twelve respectively, and the target motions were random.

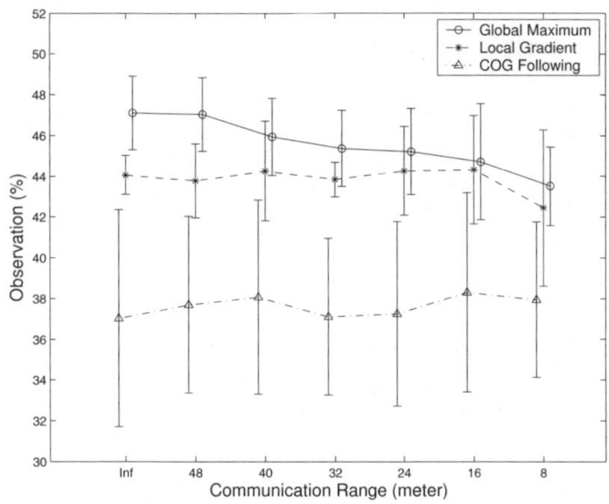

Fig. 3. Performance comparison of three coordination methods

In order to remove the effect of the underlying low-level tracker, an omni-directional, perfect sensor with 8-meter sensing range was assumed. The communication range varied from infinity to 8 meters in steps of the sensor range. Each configuration ran for 10 minutes a total of 10 times, and the average performance was taken as the final result.

Three different coordination methods were compared. The *COG Following* method controls a robot to be positioned at the center of a tracked target group so that the number of tracked targets is maximized locally. There is no communication among robots in this method. The *Local Gradient* method adopts the *Region-based Approach* with the motion strategy in Eqn. 12, which performs hill-climbing on the utility distribution function. Similarly, the *Global Max* method generates a control law based on the Eqn. 13, which controls a robot to move toward the most urgent region constantly.

The experimental results are shown in Figure 3. The *Global Max* method showed the best performance for all configurations; it was clearly shown that its performance degrades in inverse proportion to the communication range as expected. When the communication range was short, there was no significant performance difference between the *Global Max* and *Local Gradient* methods. In contrast, the effect of the communication range on the performance of the *Local Gradient* method was not noticeable except when the communication range was 8 meters. It can be understood that the *Local Gradient* method focuses more on the targets in the vicinity of a robot than those further away. Both methods outperformed the *COG Following* method, which provides evidence that coordination helps, and that the *Region-based Approach* is effective and efficient. It is also notable that the standard deviation of the *COG Following* method is larger than the other methods. This means that it is sensitive to the initial condition and the target motions. The *Region-based Approach* showed more stable performance.

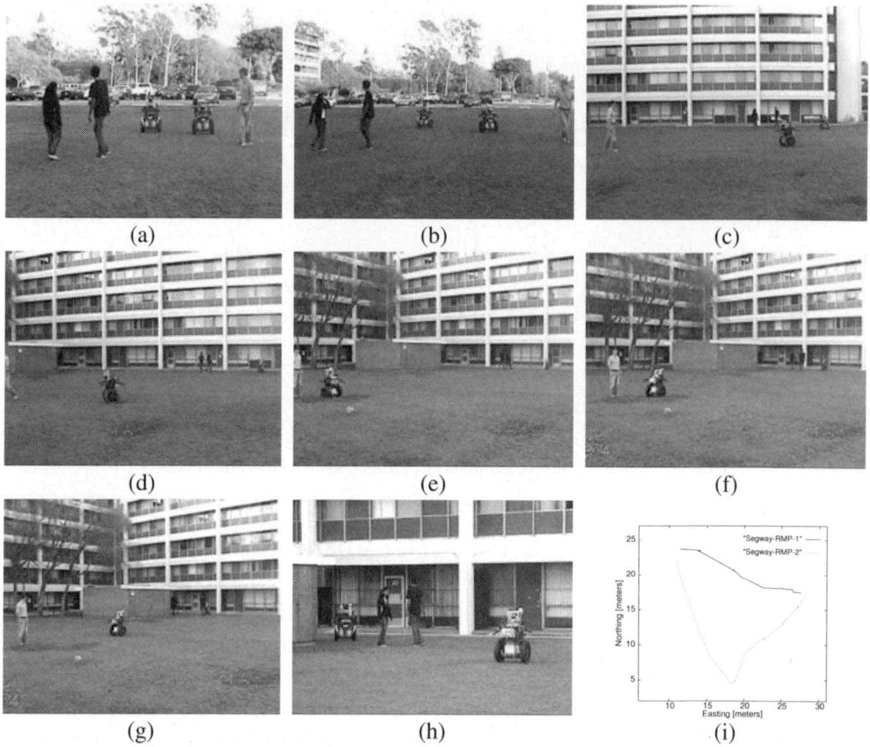

Fig. 4. Snapshots of two Segway RMP robots tracking people cooperatively

4.2 Real-Robot Experiment

The proposed algorithm was implemented and tested using real robots. The system consists of four components: target tracking, localization, cooperative motion planning, and navigation. The target trackers described in [5] were adopted for multiple target tracking. For robot localization, the data from a differential GPS and an IMU were combined using an Extended Kalman Filter. The tracking results (the target positions in a local coordinate system) and the tracker information (the robot pose in the global coordinate system) were broadcast for cooperation over a wireless network. The *Region-based Approach* described in Section 3 was utilized for cooperative motion planning. Due to limited computational power, the delta function models (Eqn. 2 and 4) and the binary model (Eqn. 6) were used to compute the utility distribution. Given its current pose (from the *Localizer* module) and the goal position (selected by the *Region-based Approach*), each robot was programmed to perform point-to-point, safe navigation using VFH+ (Vector Field Histogram +) [11].

The implemented system was tested using two Segway RMP robots. The environment was an open space (30x30 meters). The targets were three pedestrians, and they moved at regular walking speeds in the open area. The grid size of the urgency function representation was fixed to one meter. The behaviors of two robots were inspected while the number of targets changed dynamically.

The snapshots of the two Segway RMP robots tracking people cooperatively are shown in Figure 4. The robots started from the same position, and there were three people walking in front as shown Figure 4 (a). The people split into two groups as shown in Figure 4 (b): two people walking together on the left, and a single person walking in the opposite direction. As a result, two robots also split, and started to track each group respectively. Each robot broadcast its own pose and the position of tracked targets. When the single person stopped moving (Figure 4 (e)), the robot that was tracking the person lost the target and stopped. At this point, the utility value of the robot's position become low, and the robot decided to help the other robot as shown in Figure 4 (f). Finally, the robot arrived in the area whose utility value was the maximum, and helped the other robot track the targets as shown in Figure 4 (h).

The trajectory of the two robots during the experiment is shown in Figure 4 (i). The robots started from the position (28, 17). The first robot tracked the group of two people and moved to the position (11, 24). The second robot tracked the single person and moved to the position (18, 5). When the single person stopped moving, one of the peaks of the utility distribution disappeared, and the second robot moved to the position (10, 22) as a result.

5 Conclusion

In this paper, we proposed an algorithm for multi-robot coordination with applications to multiple target tracking. The proposed algorithm treats the densities of robots and targets as properties of the environment in which they are embedded, and a control law for each robot is generated by suitably manipulating these densities. Since the proposed mechanism is *on-line*, *distributed* and *expandable*, it can be applied for various sensor configurations. For example, a *heterogenous* sensor network can adopt the mechanism with minimal modification, and sensors can be added to (or subtracted from) a tracking network on the fly without stopping operation.

Two experiments has been performed to evaluate the proposed algorithm. First, two different versions of the *Region-based Approach* were compared using various configurations. Since the communication range is the most critical factor for multi-robot coordination, we varied the communication range and investigated the overall performance change. The experimental results showed that both methods outperformed the 'naive' local-following method, and it was clearly shown that our treatment of the coordination problem is effective and efficient. The developed system was also tested on two Segway RMP robots, and the behaviors of the robots in the cooperative tracking scenario provide evidence that the *Region-based Approach* controls multiple robots appropriately according to the target distribution change.

As an implementation issue, the *Region-based Approach* described in this paper can be specialized by exploiting the characteristics of an environment. For example, when the topology of the structured environment is known in advance, the representation of a utility distribution can become discrete and sparse as described in [4].

When the environment is unstructured, the grid-based representation can be adopted as described in [6].

Acknowledgment

This work is supported in part by DARPA grants DABT63-99-1-0015, and 5-39509-A (via UPenn) under the Mobile Autonomous Robot Software (MARS) program, and NSF CAREER grant IIS-0133947.

References

1. Brian Gerkey and Maja J Matarić. Principled communication for dynamic multi-robot task allocation. In D. Rus and S. Singh, editors, *Experimental Robotics*, volume LNCIS 271 of *VII*, pages 353–362, Springer-Verlag Berlin Heidelberg, 2001.
2. Ben Grocholsky, Alexei Makarenko, Tobias Kaupp, and Hugh F. Durrant-Whyte. Scalable control of decentralized sensor platforms. In *International Workshop on Information Processing in Sensor Networks*, pages 96–112, Palo Alto, CA, 2003.
3. Leonidas J. Guibas, Jean-Claude Latombe, Steven M. LaValle, David Lin, and Rajeev Motwani. A visibility-based pursuit-evasion problem. *International Journal of Computational Geometry and Applications*, 9(5):471–494, October 1997.
4. Boyoon Jung and Gaurav S. Sukhatme. Tracking targets using multiple robots: The effect of environment occlusion. *Autonomous Robots*, 13(3):191–205, 2002.
5. Boyoon Jung and Gaurav S. Sukhatme. Detecting moving objects using a single camera on a mobile robot in an outdoor environment. In *International Conference on Intelligent Autonomous Systems*, pages 980–987, The Netherlands, March 2004.
6. Boyoon Jung and Gaurav S. Sukhatme. A generalized region-based approach for multi-target tracking in outdoor environments. In *IEEE International Conference on Robotics and Automation*, pages 2189–2195, New Orleans, LA, April 2004.
7. Rafael Murrieta-Cid, Héctor González-Baños, and Benjamín Tovar. A reactive motion planner to maintain visibility of unpredictable targets. In *the Proceeding of IEEE International Conference on Robotics and Automation*, pages 4242–4247, May 2002.
8. Lynne E. Parker. Cooperative robotics for multi-target observation. *Intelligent Automation and Soft Computing, special issue on Robotics Research at Oak Ridge National Laboratory*, 5(1):5–19, 1999.
9. John Spletzer and Camillo Taylor. Dynamic sensor planning and control for optimally tracking targets. *International Journal of Robotics Research*, 22(1):7–20, January 2003.
10. Ashley Stroupe and Tucker Balch. Value-based observation with robot teams (VBORT) using probabilistic techniques. In *Proceedings of the International Conference on Advanced Robotics*, Coimbra, Portugal, June 2003.
11. Iwan Ulrich and Johann Borenstein. VFH+: Reliable obstacle avoidance for fast mobile robots. In *Proceeding of the IEEE International Conference on Robotics and Automation*, pages 1572–1577, Leuven, Belgium, May 16–21 1998.
12. Barry B. Werger and Maja J. Matarić. Broadcast of local eligibility for multi-target observation. In *Proceedings of Distributed Autonomous Robotic Systems*, pages 347–356, 2000.
13. Masfumi Yamashita, Hideki Umemoto, Ichiro Suzuki, and Tsunehiko Kameda. Searching for mobile intruders in a polygonal region by a group of mobile searchers. In *Symposium on Computational Geometry*, pages 448–450, 1997.

A Comparative Study of Market-Based and Threshold-Based Task Allocation

Nidhi Kalra[1] and Alcherio Martinoli[2]

[1] Robotics Institute, Carnegie Mellon University nkalra@cmu.edu
[2] Swarm-Intelligent Systems Group, École Polytechnique Fédérale de Lausanne alcherio.martinoli@epfl.ch

In this paper we compare the costs and benefits of market-based and threshold-based approaches to task allocation in real world conditions, where information and communication may be limited or inaccurate. We have performed extensive comparative experiments in an event-handling domain. Our results indicate that when information is accurate, market-based approaches are more efficient; when it is not, threshold-based approaches offer the same quality of allocation at a fraction of the expense. Additionally, both approaches are robust to low communication and task perception ranges in our experimental domain.

1 Introduction

Multirobot coordination has become a popular area of research and advanced significantly in recent years. Researchers have developed a wide range of coordination approaches to harness the many benefits of robot teams (including speed, robustness, flexibility, and the ability to complete a wider range of tasks) in a variety of challenging real-world application domains. Nevertheless, before multirobot systems can be used extensively, it is imperative to understand the tradeoffs between the numerous coordination schemes.

We are interested in particular in understanding the tradeoffs between self-organized approaches and intentional approaches to multirobot task allocation in real-world conditions. Self-organized approaches are fully decentralized and achieve complex collective behavior from the local interactions of many simple individuals. Robots choose their actions independently and asynchronously using positive and negative feedback mechanisms and randomness. Additionally, interactions between agents are often modulated by signs left in the environment or by local, broadcast communication. Threshold-based approaches, in particular, are popular self-organized solutions to multirobot task allocation; in such approaches, a robot's choice of activity is modulated by a perception of stimulus or demand for a task and its response threshold for that task. Alternatively, in intentional approaches, robots are typically complex and coordinate with the explicit intent of achieving a team goal. Market-based approaches, are popular intentional approaches to multirobot task allocation. In such systems, robots act as self-interested agents participating in a virtual market economy and allocate tasks by buying and selling them over the market. These approaches exploit points of centralization in the form of auctions to produce allocations. In general, self-organized approaches such as threshold-based algorithms consume fewer communication and computation resources to create a division of labor, while intentional approaches such as market-based systems consume more resources but also tend to produce more efficient allocations.

Little work has been done to truly understand the costs and benefits of exploiting communication and points of centralization for real-world task allocation. In this paper, we highlight concepts and results from our comparative study of the performances of market-based and threshold-based task allocation schemes under realistic conditions with limited communication, noisy state and task estimation, and limited task perception. We show how such conditions affect each approach and make to recommendations for using these allocation methods. Our full study is published in a technical report [1].

2 Threshold- and Market-Based Task Allocation

Given a team of robots with a common goal and a finite set of resources, the challenge of task allocation is to determine how the team's global goal be divided up and assigned to individual team members so that some global objective is met.

In threshold-based approaches, each robot has an activation threshold for each task that needs to be performed. It continuously perceives the stimulus for each task; this stimulus reflects the urgency or importance of performing that task. When a robot perceives that the stimulus for a particular task exceeds its threshold, it begins completing the task. When the stimulus falls below this threshold (e.g., when the task is completed), the agent stops executing those behaviors. This response can be deterministic or probabilistic. Threshold-based task allocation has been demonstrated in a number of domains such as foraging [2] and aggregation [3].

Alternatively, in market-based systems, robots act as self-interested agents in pursuit of individual profit. They are paid in virtual money for tasks they complete and must pay in virtual money the value of the resources they consume. Tasks typically are distributed through auctions held by an auctioneer; this auctioneer is either a supervisor agent or one of the robots. Robots compete through bidding to win those tasks that they can complete inexpensively and thus maximize their profit. This price-driven redistribution simultaneously results in better team solutions. Market-based task allocation has also been proven on a number of domains including as exploration [4] and object manipulation [5].

Prior comparisons of these approaches include work by Gerkey and Matarić [6] in which they discuss the theoretical aspects of these approaches but not their performance in realistic conditions. Campos et. al. [7] and Cicirello and Smith [8] compare the two on the problem of allocating trucks to paint booths; however, they investigate the benefits of encoding the problem as a market-based approach or a threshold-based approach, but not the costs and benefits of centralization and communication. We believe that no prior work compares these approaches along the dimensions in which we are interested.

3 Formulation of Study

In this section we describe several key components of our study, including the domain, axes and metrics for evaluation, and our implementation of the algorithms.

3.1 Event-Handling Domain

We use the event-handling domain as the framework for our study. In this domain, events occur at unpredictable times and locations throughout the environment and must be handled by the robots. In real-world scenarios, these events might correspond to machine malfunctions in a factory or requests for package pick-up in a delivery service. Respectively, handling could involve robots fixing broken machines and making deliveries. Event-handling can also be mapped onto domains that are currently solved by market-based approaches (*e.g.*, exploration [4]) and threshold-based approaches (*e.g.*, foraging [2]).

Because market-based approaches are primarily useful when task and robot state knowledge is available [1], we define event-handling precisely: robots can sense the locations of individual tasks and can estimate their own positions. The resulting task allocation problem is to assign particular events to individual robots. In some market-based approaches, a single robot may be allocated multiple tasks and keeps a schedule of these tasks. However, in threshold-based approaches, robots are almost always limited to one-task-per-robot (OTPR) allocation because there is no clear way to enable scheduling with response thresholds (this would be an interesting area of future work). Thus, in fairness, we must use OTPR allocation in the market-based approach as well. Indeed, this formulation is frequently used for simplicity in many market-based approaches (see Dias et. al. [9] for examples).

3.2 Axes, Metrics, and Experiment Parameters

We are interested in comparing the two candidate approaches under a variety of real-world conditions. Specifically, we consider the effect of poor accuracy in state estimation, poor accuracy in task localization, reduced perception range, and reduced communication range. Our goal is to compare both the quality of allocations produced and the cost of producing those allocations. We measure quality in terms of the number of events handled and the total distance traveled by the team over a fixed time window. We measure cost in terms of the computation requirements (*i.e.* algorithm complexity and empirical running time) and communication requirements (*i.e.* number of messages and the size of messages).

In all of our experiments, we use a 100×100 unit environment and run each trial to 100 iterations. There are a fixed number of robots that must handle a constant number of events; whenever an event is handled, a new one randomly appears to replace it. All our experiments were conducted using a point simulator in which robots are represented by their current locations and were run on a Linux machine with a 2.4 GHz processor and 512 Mb of memory.

3.3 Algorithms

Both market-based and threshold-based approaches have a number of parameters that can affect performance. We conducted numerous experiments to find the best formulation for each approach. Due to space considerations, we limit our discussion to resulting algorithms; full experiments can be found in our tech report [1].

In the market-based approach, our goal is to allocate an event to the robot that is closest to it. We compared two auction/bidding strategies. In the first, called M_1, we auctioned events in random order and robots bid their distances to the event being auctioned. In the second, called M_2, events were auctioned in order of increasing distance to their nearest robot. The corresponding bidding function is straightforward: a robot r without an assigned event bids a pair $\{e, d\}$, where e is r's closest unassigned event and d is its distance to e. The robot that submit the bid with minimum distance d is assigned the corresponding event. In both approaches, only robots without assignments and with information about the event being auctioned place bids, rounds of auctions are held until all events are allocated or until no bids are received, and a new round of auctions is held whenever one event is completed (this provides a reallocation method). Furthermore, robots move with perfect actuation directly to the allocated event; if no event is allocated to a robot, it does nothing until the next auction round. Our results showed that ordered auctions improve performance significantly but also consume more computation resources. In the remainder of our experiments, we use M_2 as our representative market-based algorithm. We also explored the use of a reserve price, *i.e.* the maximum value for d at which the auctioneer will award an event. This reserve keeps events from being allocated to very far away robots when all nearby robots are assigned to other events; the idea is that this event will be handled sooner if we simply wait for a nearby robot to finish with its

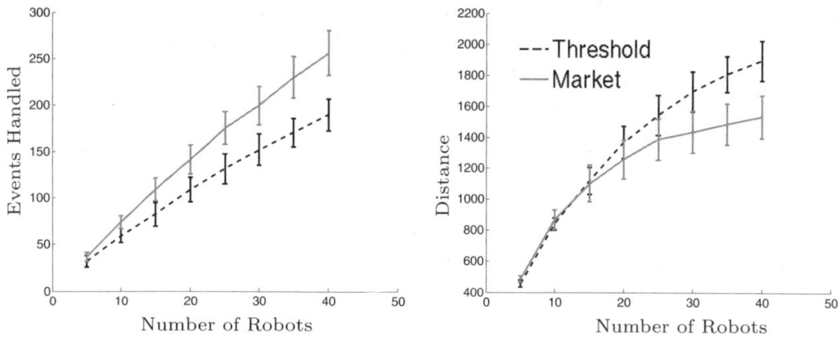

Fig. 1. Baseline comparison of the number of events handled vs. number of robots (left) and distance traveled vs. number of robots (right). We use a 100×100 environment with 20 events and compare over 100 trials with 100 iterations each. Robots have perfect task information, perfect actuation, and infinite communication range. Error bars here and throughout this paper represent one standard deviation.

current event. We found that a reserve price equal to the expected distance between robots results in the best performance.

In the threshold-based approach, our goal is for each robot to handle the event that it is closest to, without duplicating another robot's work. Thus, we formulated the stimulus $\sigma(r, e)$ produced by an event e for robot r as in Eq. 1.a, where $d(r, e)$ is the distance between the robot and the event.

$$\text{(1.a)} \quad \sigma(r,e) = \frac{1}{d(r,e)} \qquad \text{(1.b)} \quad \theta_e = \frac{1}{\langle D_r \rangle} \qquad \text{(1.c)} \quad p_e = \frac{\sigma(r,e)^n}{\sigma(r,e)^n + \theta_e^n}$$

We also explored the use of different thresholds and found that the best performance was achieved when the threshold θ_e for every event e was equal to the inverse of expected distance $\langle D_r \rangle$ between robots (Eq. 1.b). This threshold value is consistent with the best reserve price for the market. We then explored the use of deterministic versus probabilistic response. In the deterministic response, robots respond deterministically to the event e that has the largest stimulus above the threshold. In probabilistic response, they respond with probability p_e as in Eq. 1.c, where θ_e is the threshold and n is the nonlinearity of the response. We found that deterministic response outperformed all probabilistic response methods (*i.e.* for all finite values of n). We also experimented with deterministic versus random movement. We found that deterministic movement is always preferable, except when a robot is very close to its teammate and chances are high that they are handling the same event. Then, random movement for a short period of time reintroduces randomness into the system and produces better results.

To summarize, the best market-based approach uses ordered auctions and a non-zero reserve price; the best threshold-based approach uses a non-zero

Table 1. Comparing the computational and communication complexity. Communication complexity refers to the number of packets of information that must be exchanged between teammates. The size of each packet is constant. Here, r is the number of robots and e is the number of events.

Algorithm	Computational Complexity	Communication Complexity
Threshold	$O(re)$	–
Market	$O(relog(e))$	$O(re)$

threshold, deterministic response, and a mixed actuation strategy. Figure 1 presents a baseline comparison of our final threshold-based and market-based algorithm in terms of the number of events handled, the total distance traveled by the team, and the computation time required. The market-based approach always handles more events while consuming less energy traveling than the threshold-based approach; this benefit comes at the cost of significantly more computation time.

Table 1 summarizes the algorithm and communication complexity for these approaches. In the threshold-based approach, each robot computes its distance to each event once per allocation, so the complexity is $O(re)$ per allocation round, where r is the number of robots and e is the number of events. The same is true for the market-based approach; however, robots must also sort the events in order of increasing distance so they always bid on the closest event. Since most sorting algorithms run in logarithmic time of the number of items to sort, the total complexity is $O(relog(e))$, and we expect that the linear components overshadow the logarithmic component. This is supported by our empirical results as well. In terms of communication requirements, the complexity (e.g. the number of packets that must be sent) for the market-based approach is $O(re)$ because each robot sends one bid per event and the auction sends one award announcement to each robot per event. Additionally, the size of the messages sent between robots is constant. The threshold-based approach as we are currently using it has no communication requirements.

4 Comparative Results and Discussion

In this section, we explore the effects of imperfect state and task estimation, reduced communication range, and reduced perception range on market-based and threshold-based algorithms.

4.1 Imperfect State and Task Estimation

Local state estimation is imperfect in all real-world domains and usually affects the quality of allocations. To quantify the effects of poor state estimation in the event handling domain, we add Gaussian noise to each robot's position estimate during allocation. We still allow perfect actuation and perfect

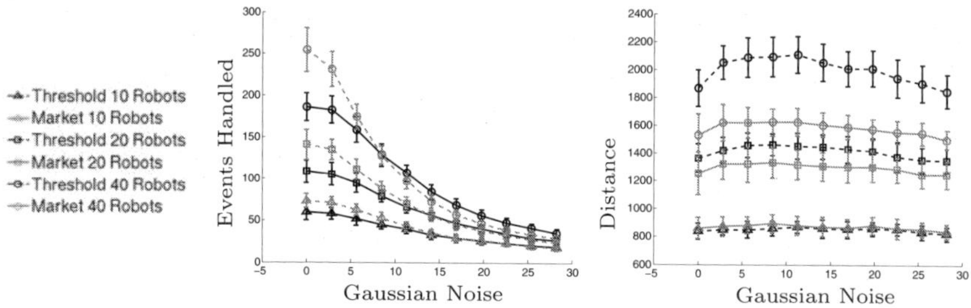

Fig. 2. Comparison of the number of events handled (left) and total distance traveled (right) vs. the standard deviation of Gaussian noise added to the robots' local position estimation. These results are almost identical to those obtained when adding the same noise to task position estimation (not shown). We use a 100×100 environment with 20 events and 10, 20, and 40 robots and compare over 100 trials with 100 iterations each.

task information to ensure that results are purely a product of localization error. This scenario is feasible if there is an overseeing agent that has access to task information and communicates that information to the robots; the robots themselves may have very poor perception.

Figure 2 plots the number of events handled by each approach (left) and the total distance traveled (right) against the standard deviation of the Gaussian noise. Qualitatively, the allocations of both algorithms degrade in the same way. Quantitatively, we see that the threshold-based approach and market-based approach have the same performance in terms of the number of events handled once the standard deviation of the error exceeds approximately 8. This ten percent error is a reality for many mobile robotic platforms, particularly smaller platforms without accurate proprioceptive sensors. The market-based approach is consistently less-costly in terms of distance when there are equally many or more robots than events. When there are fewer, the two approaches perform similarly because all robots are almost always moving to handling events.

Robots may also not have perfect task information, for instance when the supervisor or user does not have accurate information about the location of tasks but the tasks must still be allocated. In another set of experiments, we allow robots to have perfect actuation and perfect localization. This scenario is feasible if robots have poor or no task perception and the supervisor's task perception is also not perfect. However, the robots do have accurate proprioceptive sensors that provide accurate localization. Both qualitatively and quantitatively, the results (not shown) are nearly identical to the results when we compared the effect of localization error. That the effect of poor position estimation is the same regardless of whether it is the robots' or tasks' positions highlights the symmetry of our problem. That is, allocating robots to events (as the market-based approach does) or events to robots (as the threshold-based approach does) results in almost the same allocation problem.

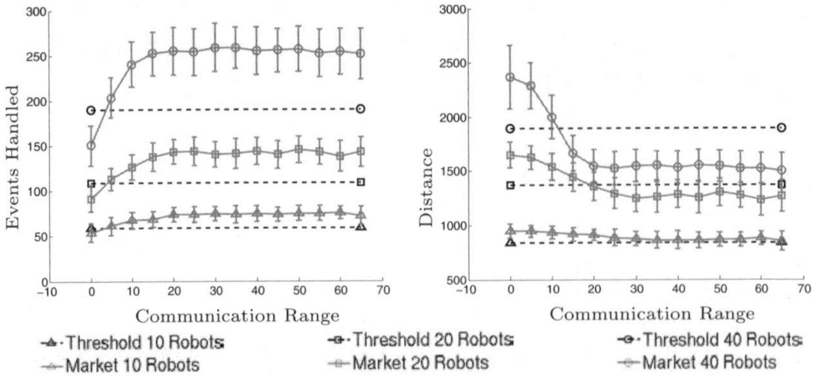

Fig. 3. Comparison of the number of events handled (left) and the total distance traveled (right),vs. the radius of the communication range. We use a 100×100 environment with 20 events and 10, 20, and 40 robots and compare over 100 trials with 100 iterations each.

Finally, when we consider these results and the cost of market-based allocations, its clear that market-based allocations are usually not worth the computation and communication cost unless accurate local state and task information is available.

4.2 Communication

Communication is essential for market-based approaches to perform auctions. However, in the real world, market-based approaches must be robust to reduced communication. Indeed, it is likely that not every member of a team will be able to communicate with every other member or with a central supervisor, if there is one. We experiment with the communication range to test the robustness of the market-based approach in this respect. Specifically, we allow auctions among only a connected set of robots. Within a connected set, we simulate a leader holding auctions for all the tasks for which it is aware. This is analogous to robots having perfect, long-range perception of the events in the environment (and no supervisor) but having limited communication range.

Figure 3 compares the performances of the market-based and threshold-based approaches with respect to the number of events handled (left) and the total distance traveled (right) by the team. First notice that the market-based approach quickly reaches its best performance in terms of events with a fairly short communication range. For 40 robots and 20 events, a range of 10-15 units suffices to produce the best results. Markets also reach their minimum total distance traveled with a fairly short communication range (about 20 units for 40 robots). The best communication range for a scenario depends on the robot and event densities. By comparing the minimum communication range required to achieve the maximum performance (in terms of events) and the size of the auctions at that range (not shown for space considerations), we found that maximum performance can be achieved with less than a third of the total number of robots participating on average in each auction. Specifically, a team of 10, 20, and 40 robots achieved maximum performance when the average auction size was about 2, 6, and, 15. This highlights that this market-based approach is fairly robust to short communication ranges.

Notice that, in this domain, the strongest candidates for a particular event are likely to be in close proximity to each other and to that event. Therefore, these candidate robots are also likely to be able to communicate with each other and thus perform an allocation of the task even when the overall communication range is low. This suggests that the robustness demonstrated here may be somewhat specific to spatial domains such as this in which spatial proximity to a teammate correlates highly with the likelihood of competing with that teammate for tasks. This important property is common to many domains including exploration, mapping, and aggregation.

Thirdly, when market-based approaches have no communication, their performance degrades significantly. When a robot cannot communicate with any teammates, it will be the only member of the communicating set and thus will hold its own auctions. However, the only bid in the auction will be its own. In effect, the robot will be moving towards its closest event, provided that closest event is within the reserve price. Without communication, the market essentially becomes the threshold-based approach, but with the added expense of internal auctions to achieve the same result.

4.3 Task Perception

Often, it is not possible for a supervising agent to be aware of the tasks in the environment, for instance in exploration or search and rescue. Thus, it is important for multirobot task allocation schemes to be able to function without *a priori* task knowledge and exploit task perception during execution.

We experimented with low task perception by varying the range at which robots can detect events. Figure 4 plots the quality of allocation versus the radius of perception. Here, we compare the threshold-based approach against three market-based approaches that assume different communication scenarios. Firstly, we assume very poor communication (range of 5), but allow robots to share perceived event information. Secondly, we allow infinite communication and again allow robots to share event information. Thirdly, we allow infinite communication but only allow robots to bid on events that they perceive on their own. These three scenarios highlight how poor perception might be improved by better communication and information sharing.

Notice that again the minimum perception range required to achieve maximum performance is quite short, a range of 20 suffices for all approaches under all conditions. We believe this robustness is also somewhat specific to spatial domains such as this. In the threshold-based approach, as long as a robot

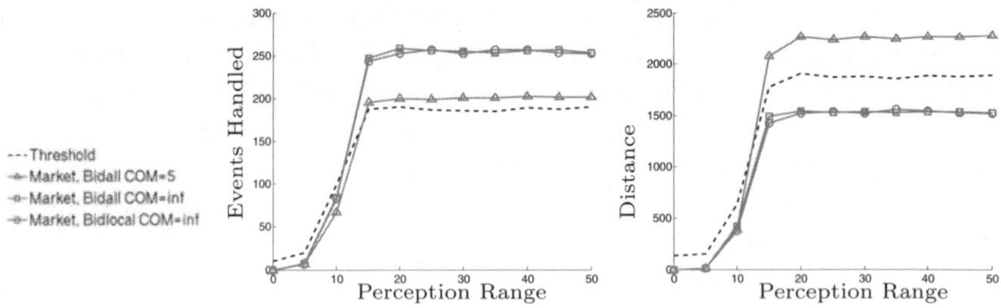

Fig. 4. Comparison of the number of events handled (left) and the total distance traveled (right) vs. the radius of the perception range, for the threshold based approach and three variants of the market based approach: one with poor communication and local bidding, one with infinite communication and local bidding, and one with infinite communication and global bidding.

can perceive its single closest event, it can begin handling that event. For the market, there is additional benefit in perceiving other nearby events in case a robot does not win the auction for its single closest event. In these experiments, a robot's expected distance to its closest event is approximately 12. Thus, significantly greater perception range is not required for sufficient performance. This conclusion is further supported by the fact that performance is not improved in the market-based approach by robots sharing information about perceived tasks and bidding on all events rather than just bidding on locally perceived events. The idea is that if a robot did not perceive an event, it is probably not the best candidate to handle that event. Instead, we attribute the difference between the market-based approaches seen here to the increased communication range.

5 Conclusions

In this paper we have compared the performances of a self-organized and an intentional approach to multi-robot task allocation. Specifically, we compare the costs and benefits of exploiting points of centralization as done in market-based approaches to fully distributed threshold-based approaches in real world conditions. We evaluate how accuracy in local state and task estimation, communication range, and task perception range affect the quality of allocations on an event handling domain.

Our results indicate that when information is accurate, market-based approaches are more efficient (though threshold-based approaches with communication must still be considered). When information about tasks and local state is not accurate, market-based approaches may not be worth the added expense; rather, threshold-based approaches offer the same quality of allocation at a fraction of the expense. Additionally, both approaches are robust to low communication and task perception ranges. We hypothesize that this is in part due to the spatiality of our experimental domain, a property that is common to a number of real-world domains. Furthermore, although our experiments use specific instances of these algorithms, we believe the results can be generalized to other variations since many of those we explored (e.g. probabilistic response thresholds) [1] follow similar trends.

In the near future, we hope to verify the results of our microscopic simulations with a realistic simulator. In the longer term, we hope to perform experiments with real robots.

Acknowledgments

This work was carried out at EPFL where Nidhi Kalra was sponsored by an EPFL graduate student fellowship. Alcherio Martinoli is funded by a Swiss National Science Foundation Grant, contract No. PP002-68647.

References

1. N. Kalra and A. Martinoli, "A comparative study of market-based and threshold-based multirobot task allocation," EPFL, Lausanne, Switzerland, Tech. Rep. SWIS-IP1, February 2006.
2. M. J. B. Krieger and J.-B. Billeter, "The call of duty: Self-organized task allocation in a population of up to twelve mobile robots," *Robotics and Autonomous Systems*, vol. 30, no. 1-2, pp. 65–84, 2000.
3. W. Agassounon and A. Martinoli, "Efficiency and robustness of threshold-based distributed allocation algorithms in multi-agent systems," in *Proceedings of the International Joint Conference on Autonomous Agents and Multiagent Systems*, 2002, pp. 1090–1097.
4. R. M. Zlot, A. Stentz, M. B. Dias, and S. Thayer, "Market-driven multi-robot exploration," Robotics Institute, Carnegie Mellon University, Pittsburgh, PA, Tech. Rep. CMU-RI-TR-02-02, January 2002.
5. B. P. Gerkey and M. J. Matarić, "Sold!: Auction methods for multi-robot coordination," *IEEE Transactions on Robotics and Automation*, vol. 18, no. 5, pp. 758–768, Oct. 2002.
6. B. P. Gerkey and M. J. Mataric, "A formal analysis and taxonomy of task allocation in multi-robot systems," *International Journal of Robotics Research*, vol. 23, no. 9, pp. 939–954, 2004.
7. M. Campos, E. Bonabeau, G. Theraulaz, and J.-L. Deneubourg, "Dynamic scheduling and division of labor in social insects," *Adaptive Behavior*, vol. 8, no. 2, pp. 83–94, 2001.
8. V. Cicirello and S. Smith, "Wasp-like agents for distributed factory coordination," *Journal of Autonomous Agents and Multi-Agent Systems*, vol. 8, no. 3, pp. 237–266, 2005.
9. M. B. Dias, N. Kalra, R. Zlot, and A. Stentz, "Market-based multirobot coordination: A survey and analysis," *IEEE Special Issue on Multirobot Systems*, 2006 (forthcoming).

Single Operator, Multiple Robots: Call-Request Handling in Tight-Coordination Tasks*

Gal A. Kaminka and Yehuda Elmaliach

The MAVERICK Group, Computer Science Department
Bar Ilan University, Israel
{galk,elmaley}@cs.biu.ac.il

Summary. Many applications of robots require a human operator to supervise and operate multiple robots. In particular, the operator may be required to resolve *call requests* when robots require assistance. Previous investigations assume that robots are independent of each other, and allow the operator to resolve one request at a time. However, key challenges and opportunities arise when robots work in tightly-coordinating teams. Robots depend on each other, and thus a single failing robot may cause multiple call requests to be issued (by different robots). Moreover, when the operator switches control to a robot, its teammates must often wait idly until the call request is resolved. We contrast previous approaches with two novel *distributed* methods, where the call-request resolution is itself considered a collaborative problem-solving activity, and non-failing robots use their knowledge of the coordination to assist the operator. We empirically compare the different approaches in several scenarios involving tight coordination, where an operator seeks a dead robot in order to assist it. Extensive experiments with 25 human operators show that this new technique is superior to existing methods, in terms of reducing the time to locate the dead robot. We also show that the new method has much more consistent performance across different operators.

1 Introduction

There is need for human intervention in applications of robot teams. Teams of multiple robots can carry out mundane or dangerous tasks. However, many applications require occasional human intervention, either for safety reasons, or because the robots suffer from some failure that requires resolution by the operator. Examples of such applications include search and rescue operations [7], multi-rover planetary exploration, and multi-vehicle operations [4].

Previous work has examined centralized methods in which a single operator interacts with multiple robots, both for monitoring their activity, as well as for resolving contingencies [1,3,4,8,11]. Here, robots are assumed to operate autonomously, as direct teleoperation of all robots in parallel is impractical. Robots that require the operator's assistance initiate or are issued *call-requests*, which are queued for the operator. The operator switches control between robots, and uses single-robot teleoperation with individual robots to resolve the call requests in some (prioritized)

* This research was supported in part by BSF Grant #2002401 and by the Israel Ministry of Science and Technology (MOST).

sequence. This method works well in settings where the task of each robot is independent of its peers, and thus the resolution of call requests can be done in sequence, independently of other call-requests.

Unfortunately, these centralized methods face difficulties in *coordinated tasks*—tasks that require tight, continuous, coordination between the robots, i.e., robot teams where robots are highly inter-dependent. First, due to the coordinated nature of the task, robots depend on each other's execution of subtasks; thus a single point of failure (e.g., a stuck robot) will quickly lead to multiple call requests. Second, when the operator switches control to a robot, the other robots must wait for the resolution of the call-request, because their own decision-making depends on the results of the operator's intervention. As a result, robots wait idly while the call request is resolved. While monitoring and diagnosis techniques can help localize call-requests to the relevant robot [5], minimizing the duration of call-request resolution remains a key challenge.

Operating a team of coordinated robots raises the opportunity for novel resolution methods, in which the responsibility for the resolution of the call request is *distributed*. Rather than having the operator centrally take all actions to resolve a failure, the otherwise-idle robot teammates can offer assistance, e.g., in providing useful information or in carrying out sub-tasks associated with the resolution process.

For example, consider the task of controlling three robots moving in formation (a task requiring tight continuous coordination between robots). Suppose one of the robots gets stuck, and is unable to move. A call request is issued to the operator, which must identify the failure and attempt to resolve it in some fashion. Previous approaches would have the operator attempt to teleoperate the robot in an attempt to dislodge it, while the other robots are idle [1,4].

However, the operator could take advantage of the other robots to resolve the failure. First, the other robots could be used to provide video imagery of the stuck robot from various angles. Second, the robots may assist the operator to determine the location of the robots—since they can calculate its expected position with respect to their own position—based on its position within the formation.

This paper takes first steps towards allowing robots to use their knowledge of the coordination to autonomously assist the operator. We examine several variations of a distributed control methodology in which functioning members of the team, rather than switching to an idle mode of operation, actively seek to assist the operator in determining the failure. The key idea is that the responsibility for resolving the call-request is distributed among the team-members *in addition to* the operator.

We empirically evaluate these variations (and contrast them with previous approaches) in extensive experiments with 25 human operators, operating a team of 3 Sony AIBO robots. The experiments evaluate several concrete call-request scenarios, in which a stuck robot must be located by the operator. The results show that distributed call-request resolution leads to shorter failure-recovery times. Moreover, the results show that a key factor in the success of the distributed method lies in the robots' use of organizational knowledge (i.e., their knowledge of the coordination). However, even in cases where this organizational knowledge fails, the operator is able to compensate. Thus the use of our distributed approach is always better than either the operator or the robots resolving the call request by themselves. A final promising result is that the distributed methods lead to improved operator consistency, reducing the variance in performance between operators.

2 Background and Motivation

The bulk of existing work on controlling multiple robots put the operator in a centralized role in attending to robots, and do not often distinguish between different task types on the basis of the coordination involved. Indeed, many existing approaches implicitly assume that robots are relatively independent in their execution of subtasks. As a result, a centralized control scheme does not interfere with task execution. Fields [3] discusses unplanned interactions between a human and multiple robots in battlefield settings, where otherwise-autonomous robots send call requests to the human operator to ask for assistance. These call requests are queued, and the operator resolves the problems one by one. Fong et al. [4] propose a *collaborative control* system that allows robots to individually initiate and engage in dialog with the human operators. The call requests are queued based on priority, and resolved serially.

Myers and Morely [8] discusses an architecture called TIGER that uses a coordinating agent that mediates between the operator and autonomous software agents. This agent centralizes the information from all agents, and can present it to the operator (or provide it to other agents). The agent is also responsible for translating operators instructions to the team. This approach thus assumes that call requests may be resolved autonomously by the robots, given appropriate high-level commands to the team. In contrast to this approach, we believe that often, the operator must directly interact with a failing robot or its teammates to resolve a call request. We thus allow the operator to directly interact with any single robot, while others assist.

ACTRESS (Actor-based Robots and Equipment Synthesis System) [10] is a multi-agent robot architecture which incorporates an interface for monitoring and controlling robots. The operator may issue commands that affect groups or individual robots. However, ACTRESS does not utilize collaboration between the operator and robots in resolving call requests. The operator may issue commands to robots that assist in such resolution, but the robots are otherwise idle.

In contrast to the above centralized approaches, we believe that resolving call-requests is in the interests of all robots currently coordinating with the robot requiring assistance—and thus they should actively collaborate with the operator to resolve the call request. Other work has also examined distributed paradigms for human/robot interaction. Tews et al. [11] describe a scalable client/server architecture that allows multiple robots and humans to queue service requests for one another. Scerri et al. [9] describe an architecture facilitating teamwork of humans, agents and robots, by providing each member of the team with a proxy and have the proxies act together as a team. Our work differs from both of these investigations in that we do not attempt to put humans and robots on equal ground. In our current work, only the human can initiate the distribution of a task to resolve a call-request. However, once initiated, the task is carried out by all members of the robotic team and the operator.

Ali [2] compares different classes of human-robot team interaction (*Direct manual control, supervisor control, individual* and *group control*). The parameters measured are effectiveness (in term of task completion and speed of completion), safety (both for the robots and their environment), and ease of use. While we similarly evaluate different interaction methods, we focus only on the case of one operator and multiple robots. However, within those, we distinguish several different types. Moreover, we provide new distributed resolution types.

3 Distributed Call-Request Resolution

As previously discussed, centralized resolution of call requests, by the operator, may work well when robots' tasks are independent of each other. However, in coordinated tasks, many robots may have to stop their task execution until a call request is resolved, because their own task execution depends on that of the robot that requires the resolution. In such cases, it is critical to minimize the time it takes to resolve a call request.

We thus focus on a distributed control approach, whereby the robots who depend on the resolution of the call-request take active steps to resolve it, in collaboration with the operator. This approach takes advantage of the robot teamwork, by turning the resolution of the call-request into a distributed collaborative task for all involved. Moreover, the active robots (that do not require assistance) are involved in a coordinated effort with the robot requiring assistance, and thus may be in a better position to assist it.

The key idea behind this approach is that call-request resolution is best viewed as an instance of cooperative problem-solving. During task execution, robots collaborate to achieve the operator goal. If task execution is halted due to a failure, a new collaboration problem instance is generated (resolving the call-request), which should then be addressed by the team-members that are affected by the failure, since they have knowledge which they can bring to bear on the problem.

Concretely, we investigate distributed resolution in repairing broken formations of Sony AIBO 4-legged robots. Formation-maintenance tasks require tight, continuous coordination between robots [6]. When a robot fails and is unable to move, the formation cannot proceed until the failure is resolved in some fashion: Either the robot becomes unstuck, or it is declared dead and the formation proceeds without it. A stuck robot often cannot report on why it is stuck, due to sensory range limitations. For instance, in the AIBO robots, the camera (mounted in the head) cannot pan and tilt to cover the rear legs. Thus if one of them is caught by something, the robots own sensors cannot identify it. The robot must then issue a call-request for assistance. The operator, in turn, must use one of the other robots to locate the stuck robot and get video imagery of its state. This act of locating the other robot and getting sufficiently close to it is a key factor in the resolution of the call request in this case.

We contrast different resolution schemes. The first, *teleoperated* scheme corresponds to the centralized control used in previous approaches (e.g., [1, 4]). In this scheme, the operator would switch control from one active robot to the next, as deemed necessary, and manually teleoperated controlled robots (one at a time) until the disabled robot was found. When one robot is controlled, the others remain idle. Another previously-investigated approach is the fully *autonomous* scheme, that lets the active robots (but not the operator) search for the failing robot. This scheme corresponds roughly to the method described in [8], where the robots receive general instructions (here, "search!") by the operator, but are left to translate and follow these commands autonomously, without direct manipulation.

We compare these previous approaches to two variations of distributed call-request resolution. In the first (*semi-distributed*), the robots assist the operator by autonomously beginning to search for the failing robot as soon as the call request is received. The operator views a split-screen view of their video imagery, and as soon as it identifies the stuck robot in one of the displays, can switch control to the robot associated with the display. Once a robot is taken over by the operator, the others become idle. The operator may still switch control to these other robots, but they no longer work in an autonomous fashion.

The fully-distributed scheme mixes teleoperation and autonomous search all through the call resolution process. The operator may teleoperate any robot at any time, and may switch between controlled robots as needed. When not operator-controlled, the robots first head towards the expected position of their stuck peer. This position is estimated based on their knowledge of the formation (organizational knowledge), under the assumption that the robot became stuck in its previous location within the formation. If they fail to find it there, they begin a spiral search pattern that is likely to find the robot, but may take relatively long time. Thus the operator and the other robots work in parallel: The search ends when either an autonomous robot or a teleoperated robot are sufficiently near the stuck robot.

The motivation for the distributed scheme is that the robots may be able to use their knowledge of the robot's role in the formation to attempt to locate it. The robots that maintain the formation have improved chances to localize themselves with respect to the formation, then an operator which takes control of a robot in the formation, without the situational awareness of the robots. On the other hand, the operator has superior inference and vision, and may be able to locate the stuck robot in the video imagery, even in cases where the robots would be unable to do it.

4 Experimental Evaluation of Call-Request Resolution Methods

We now turn to empirically evaluate these call-request resolution schemes with human operators. We use all schemes in failure scenarios in the context of a triangular formation of three robots. In each of the failure cases, we disable one of the robots to simulate a catastrophic failure, not letting it move or communicate. In accordance with previous approaches, a call-request is then issued to determine the whereabouts of the failing robot. The robots and operator then begin the search process as described above. The search stops when any robot is within a predetermined distance of its failing teammate.

We examine three scenarios. In all, the right follower robot was disabled, and color marked to allow its detection by the other robots (called *active robots*) and the operator. A potential advantage of the distributed and autonomous schemes is that they can utilize the robots' own knowledge of the coordination to locate the stuck robot. In particular, because the robots have moved in formation prior to the call-request, they may have an easier time guessing their peer's location than the operator (who needs to orient herself in space via the teleoperated camera).

To evaluate the importance of this advantage, we varied the position of the disabled robot (Figures 1, 2): The *easy* setup placed the disabled robot at approximately where it would be had it just stopped in its tracks prior to the team getting notification of the call request, i.e., a bit farther behind its location within the formation (Figures 1-a, 2-a). The *medium* setup placed the robot behind the left follower robot (1-b, 2-b). The *difficult* setup placed the robot to the left of the left follower robot, and behind it, i.e., completely out of place compared to the formation (1-c, 2-c). Thus the locations progress from a location easily predictable by the robots, to a location unpredictable to them.

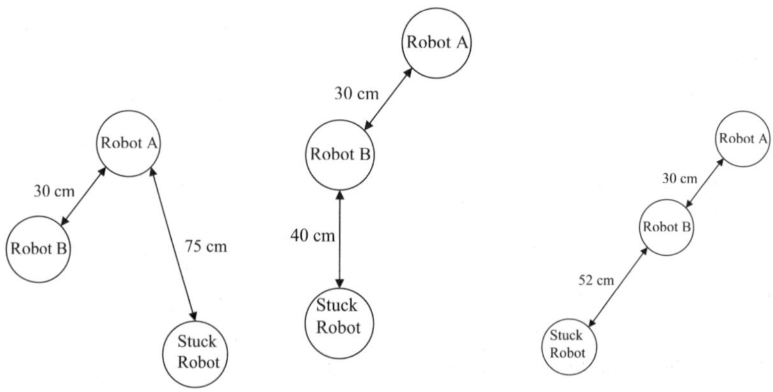

(a) Easy (b) Medium (c) Diffi cult

Fig. 1. The three experimental setups, distinguishing predictable, semi-predictable, and unpredictable locations of the stuck robot.

(a) Easy (b) Medium (c) Diffi cult

Fig. 2. AIBO robots in initial places for the three experimental setups.

We tested 25 human operators with each of the failure scenarios (22 male, 2 female; 22 of these—including the two females—were graduate or undergraduate students). All operators were novices; none had previous experience controlling multiple robots. Each operator tried all the the resolution schemes previously described, in each of the three scenarios. The ordering of the scenarios was randomized between operators to prevent biasing the results.

We distinguished two phases: The first phase of the resolution involved recognition of the disabled failure from any distance. The second phase involved its localization by another robot reaching within 35 centimeters of it. Each scenario began with the simulated disabling of the robot (and issuing of the call request), and ended with its localization by at least one robot—teleoperated or autonomous.

For each of the failure scenarios and for each method, we measure the duration of the two phases. This is an objective performance measure because the initial locations of the robots are fixed, the searching speed is constant for non-teleoperated robots, and the termination condition for the search are fixed (robots within specific distance of the failing robot). Thus other than the typical robot sensor uncertainty, performance variance is introduced solely by operator intervention. The first measured duration is that of the time that it took the operator to recognize the disabled robot in any one of the cameras (the operator uses the split-view interface in this task), i.e., the duration of the first phase. In all but the teleoperated scheme, the operator is completely passive during this interval. We then measure the time that it takes for an active robot—autonomous or teleoperated—to reach the disabled robot, i.e., the duration of the second phase. Since the motivation behind the distributed control scheme is to reduce the time spent awaiting resolution, we prefer shorter overall durations.

We begin by examining the bottom line—the total time it takes to identify the location of the disabled robot. Figure 3 shows the average total duration for the 25 operators. The vertical axis measures the time in seconds, while the horizontal axis shows the three experiment setups. In each, four bars are shown corresponding to the different resolution schemes (left-to-right: Autonomous, semi-distributed, distributed, and teleoperated).

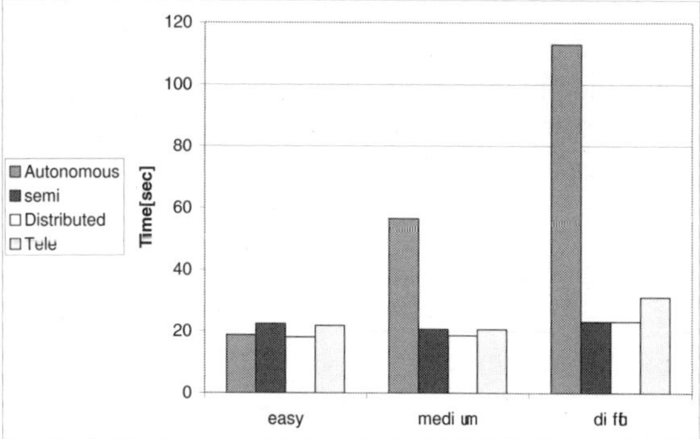

Fig. 3. Total Time to Resolution (in seconds).

The results show that in all *easy*, *medium* and *difficult* locations, the distributed approach is preferable to the both centralized teleoperation approaches, and the fully autonomous approach. Full distributed search does better than the semi-distributed approach in all locations, and better or same than the autonomous approach or same.

Overall, the distributed collaboration between the operator and active robots in the distributed approach proves to be a powerful technique for significantly reducing the time to complete the task of locating the disabled robot.

The results have been tested using a one-tailed t-test assuming unequal variances. In the easy setup, the distributed scheme is not significantly different than the autonomous scheme, and only moderately different ($p < 0.12$) than the semi-distributed and teleoperated schemes. However, as we move to the medium and difficult setups, the situation changes. The total time for the distributed scheme is significantly lower than the total time for the autonomous scheme in the latter setups ($p < 0.00004$ and $p < 10^{-12}$, resp.). The distributed scheme does better than the teleoperated scheme in the difficult setup ($p < 0.02$), and is moderately better in the medium setup ($p < 0.13$).

The figure also carries other lessons. First, the ability of the robots to use organizational knowledge of the formation can be very useful in reducing the resolution time, and thus in assisting the operator. When the stuck robot was located approximately where it was predicted to be in terms of its position in the formations, the robots were able to quickly locate it, in fact beating the operator in terms of total time (see more on this below). However, the distributed scheme was superior even in these cases, because even in where the robots were not as successful, the operator (working in collaboration with the robots) was able to compensate. This is particularly evident as the difficulty of the different setups increased, and the location of the stuck robot was unpredictable to the robots.

To better understand these results, we should consider separately the results for the first phase of the search (when an remote identification of the stuck robot was made by the operator), from the second phase, in which an active robot was to approach the stuck robot to localize it. Figure 4 shows the results of the different control schemes for the first phase, averaged across operators. The figure measures the average time (in seconds) it took the operator to recognize the disabled robot from afar, in the split-view camera display. In the autonomous approach, the operator did not intervene in the operation of the robots, only indicated that the stuck robot was recognized. In the teleoperated scheme, the operator manually turned a robot around until a heading to the remote robot was recognized.

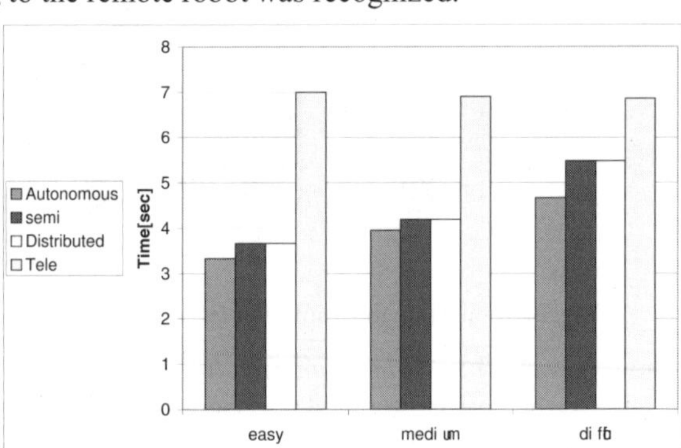

Fig. 4. Phase 1 *Time until initial (remote) identifi cation (in seconds).*

Clearly, all approaches in which robots attempt to orient themselves towards the predicted location of the disabled are superior to a teleoperated (centralized) approach. Note that in all approaches, the operator recognizes the robot from afar. The active robots do not necessarily recognize the other robot from afar, and as we will see below, may end up searching for it in the wrong location. This significantly shorter initial recognition is a beneficial side-effect of the distributed approaches. However, the initial benefits of the robots to orient themselves towards the stuck robot is lost in more difficult settings.

Figure 4 also shows an important property of the usefulness of human operators: Human ability to recognize the robot from afar is virtually identical in all three difficulty settings. Thus humans bring to bear consistent robust (if slow) capabilities. These can be useful in real applications, where the stuck robot may be partially hidden behind obstacles or otherwise not visible at all to the robots.

An examination of the second phase of the search (once an approximate heading towards the stuck robot is determined) is also telling with respect to this issue. Figure 5 shows the results for this phase, where the task is to arrive within the proximity of the disabled robot. Despite its poor performance in phase 1, the teleoperated approach does quite well in phase 2. This is easily explained—here the disabled robot is already recognized, and the teleoperated approach simply allows the operator to now drive the teleoperated robot as quickly as possible, outrunning automatic approaches that move in constant (and typically conservative) speed. Thus again, the operator brings to bear capabilities that cannot be duplicated by the robots.

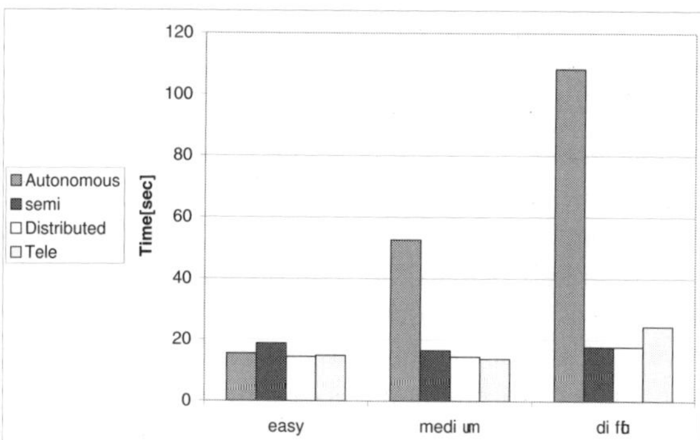

Fig. 5. Phase 2 *From initial identifi cation to localization of the stuck robot (in seconds).*

However, the best performances was by the distributed approach, because it essentially turns this phase into a race between a teleoperated robot and an autonomous robot, as to who gets to the disabled robot first. Moreover, unlike the semi-distributed approach, where there's an overhead of a few seconds while the operator takes over control (see the results for the easy/medium location), here the transition from phase 1 to phase 2 is fairly smooth, because one active robot continues to search even while the operator is taking over control of the other. Thus there is here a composition between the *Autonomous* approach and the *Teleoperated* approach.

Indeed, contrasting the results of the *Autonomous* and *Distributed* approaches is telling. As we move from the easy location to medium to difficult, the gap between the methods is grows in favor to the *Distributed* approach. That happens as a result of the inability of the *Autonomous* approach, to locate the stuck robot in unpredictable places. The collaboration between the human operator and the robot team is superior to either, alone.

An final lesson is revealed by examination of the standard deviation of the results for total task-completion time. Table 1 shows the standard deviation for the different approaches, in the three experiment setups. Each row corresponds to a different method, and each column to different setup. We can see that in the easy setup, the autonomous, semi-distributed, and distributed schemes all have essentially the same standard deviation, indicating similar performance. However, the standard deviation for the autonomous scheme in the medium setup is much higher than for the other approaches. In the hard setup, both the autonomous and teleoperated approaches suffer from greater standard deviation in performance than the two distributed schemes. This shows an additional benefit of the distributed methods: A more consistent performance of operators in the distributed and semi-distributed cases.

	Easy	Medium	Diffi cult
Autonomous	11.21	34.64	23.82
Semi-Dist.	11.30	5.07	7.78
Distributed	11.29	5.16	7.90
Teleoperated	7.68	5.96	15.87

Table 1. Standard deviation of call-resolution times (in seconds).

5 Summary and Future Work

This paper explores novel first steps towards distributed call-request resolution schemes, in which the operator and robots collaborate to resolve failures. This scheme is particularly suited to situations where robots are tightly coordinated, and thus are able to use their knowledge of the coordination to effectively assist the operator. The technique builds on a key idea, that the resolution of failures in cooperative tasks should be viewed as a cooperative task in itself. Previous techniques (teleoperation of one robot at a time, autonomous operation of the robots) were meant for tasks that do not require tight coordination between the robots.

We empirically evaluate this new technique and compare it to previous work, in extensive experiments with 25 human operators. The results show that the distributed control scheme, exploiting teamwork between the operator and all robots, reduces the time of resolving failures (compared to the centralized and autonomous approaches), and was superior in all cases. Moreover, the technique leads to reduced variance between operators. However, its overall improvements with respect to the centralized teleoperated approach was only significant in a subset of the experimental conditions.

The promising results presented in the paper also raise important questions for future work. We are particularly interested in integrating the new technique with complete human-robot interaction systems, in order to evaluate its effectiveness not only within call-request resolution, but in more general settings of operating the robots even in non-failures cases.

Acknowledgments

We thank the anonymous reviewers and Avi Rosenfeld for their helpful comments. Thanks to K. Ushi.

References

1. J. A. Adams. *Human Management of a Hierarchical System for the Control of Multiple Mobile Robots*. PhD thesis, University of Pennsylvania, 1995.
2. K. S. Ali. *Multiagent telerobotics: Matching systems to tasks*. PhD thesis, Georgia Institute of Technology, 1999.
3. M. Fields. Modeling the human/robot interaction in onesaf. In *Proceedings of the 23rd Army Science Conference*, 2002. (Poster).
4. T. Fong, C. Thorpe, and C. Baur. Multi-robot remote driving with collaborative control. *IEEE Transactions on Industrial Electronics*, 50(4):699–704, August 2003.
5. G. A. Kaminka and Y. Elmaliach. Experiments with an ecological interface for monitoring tightly-coordinated robot teams. In *Proceedings of IEEE International Conference on Robotics and Automation (ICRA-06)*, 2006.
6. G. A. Kaminka and R. Glick. Towards robust multi-robot formations. In *Proceedings of IEEE International Conference on Robotics and Automation (ICRA-06)*, 2006.
7. R. Murphy, J. Casper, M. J. Micire, and J. Hyams. Mixed-initiative control of multiple hetrogeneous robots for urban search and rescue. Technical Report CRASAR-TR2000-11, Center for Robot-Assisted Search & Rescue, University of Southern Florida, 2000.
8. K. L. Myers and D. N. Morely. Human directability of agents. In *Proceedings of the First International Conference on Knowledge Capture, K-CAP 2001*, Canada, 2001.
9. P. Scerri, L. Johnson, D. Pynadath, P. Rosenbloom, M. Si, N. Schurr, and M. Tambe. A prototype infrastructure for distributed robot, agent, person teams. In *AAMAS-03*, 2003.
10. T. Suzuki, K. Yokota, H. Asama, H. Kaetsu, and I. Endo. Cooperation between the human operator and the multi-agent robotic system: Evaluation of agent monitoring methods for the human interface system. In *Proc. of the 1995 IEEE/RSJ International conference on intelligent robots and systems*, pages 206–211, 1995.
11. A. D. Tews, M. J. Mataric, and G. S. Sukhatme. A scalable approach to human-robot interaction. In *ICRA-03*, 2003.

Distributed Metamorphosis Control of a Modular Robotic System M-TRAN

Haruhisa Kurokawa[1], Kohji Tomita[1], Akiya Kamimura[1], Satoshi Murata[2], Yuzuru Terada[2], and Shigeru Kokaji[1]

[1] National Institute of Advanced Industrial Science and Technology (AIST)
{kurokawa-h,k.tomita,kamimura.a,s.kokaji}@aist.go.jp
[2] Interdisciplinary Graduate School of Science and Engineering, Tokyo Institute of Technology
murata@dis.titech.ac.jp, string@mrt.dis.titech.ac.jp

Abstract. Metamorphosis by a self-reconfigurable modular robot is presented in this report. We have developed a new prototype, "M-TRAN III", which is improved in its high speed and rigid connection mechanism. Using its integrated design of a multi-CPU controller with various programming tools, experiments of self-reconfiguration were successfully carried out through single master synchronous control. Based on the obtained results, decentralized and locally synchronous control was accomplished, which controlled self-reconfiguration of up to 20 modules using the same program.

1 Introduction

A modular robot contains numerous autonomous modules, each of which has actuators, sensors, and processors. Modules are able to connect mechanically with each other and communicate and cooperate with others. Even with few degrees of freedom in connection and motion of a single module, the total system can form various structures and make flexible motions. Self-reconfigurable modular robots can control mechanical connections among modules by themselves and perform reconfiguration.

A modular robot, as a distributed autonomous system, is expected to be flexible, robust, fault-tolerant, and adaptive, as in the case of living creatures, if various functions of the total system, such as morphology and motion, are produced through self-organization from a homogeneous entirety. By using modular robots, coevolution/adaptation of morphology and motion, both at the same time, can be studied. To date, hardware studies of modular robots are dealing with two subjects, locomotion and self-reconfiguration. This paper specifically addresses self-reconfiguration through distributed control, as an attempt toward a distributed autonomous system.

After pioneering studies of CEBOT [1], two-dimensional systems "Fracta" [2] and "Metamorphic Robot" [3], were developed. In both studies, metamorphoses from an arbitrary configuration to a target configuration were attempted through distributed control. Subsequently, several three-dimensional robotic modules [4–8] and various algorithms [9–11] were proposed.

Development of a three-dimensional system remains challenging. Hardware performance is largely dependent on mechanical design. Although various robots of different geometry and mechanisms have been developed, most have realized only small-scale self-reconfiguration. Algorithms of large-scale self-reconfiguration have been verified only by simulation. Difficulties of experimental verification have been attributed to these limitations: 1) few modules were produced, 2) modules lacked sufficient strength to manipulate or hold the structure, and 3) connections were less reliable, too slow, or too power-consuming.

We have continued development of systems called Modular Transformers (M-TRANs) [12]. Using two prototypes, M-TRAN I & II, various 3-D self-reconfigurations and distributed control of locomotion were achieved [12–14]. However, the modules in the experiments remained numerically limited; complicated reconfiguration was not possible mostly because of the third reason explained above.

To overcome the problems described above, we developed a third prototype, whose connection mechanism is faster and less power consuming. We developed an onboard distributed controller, designed geometric sequences of metamorphoses, and verified the performance of the system through various experiments of self-reconfiguration. In all experiments, the modules were set the same: the total system was homogeneous. After verification of hardware and controller systems, experiments using decentralized and locally synchronous control progressed.

In the following section, the basic design of the M-TRAN module and sequences of metamorphosis are described to define the research objectives. Hardware and software development are detailed in Sections 3 and 4. Experimental results are described in Section 5. Section 6 concludes the paper and suggests subjects for future studies.

2 M-TRAN Module Design and Self-Reconfiguration

2.1 Basic Design

An M-TRAN module comprises two blocks and a link (Fig. 1 (a))[12]. Both blocks are identically shaped; they are made of a half-cube and a half-cylinder with no protrusions. Two blocks can rotate independently about their axes by ±90 degrees. At any angle, two half-cylindrical parts always contact with each other.

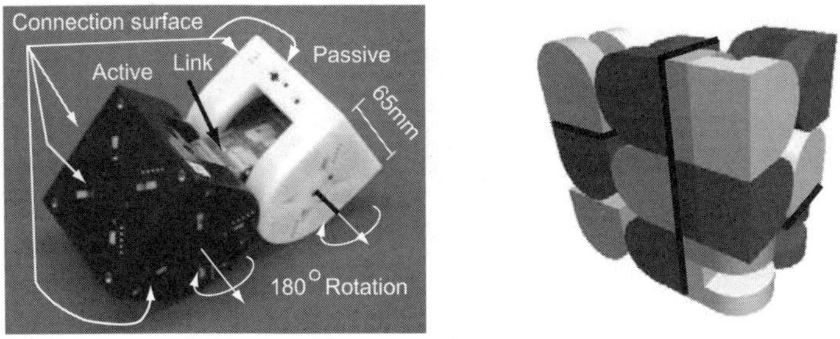

(a) M-TRAN module

(b) M-TRAN in connection

Fig. 1. Basic design of M-TRAN module.

Each block has three connection surfaces. Although all three are not of the same shape – one is square and two are half-rounded – they all have the same connection function. The two blocks differ in gender and a connection surface of an active (male) block can connect mechanically with a passive (female) surface of another module among four possible orientations. Using that connection mechanism, modules can be assembled to form various configurations. Moreover, each module can control its connection and change its total configuration. Self-reconfiguration is conducted mainly by controlling angles of all modules in 0 or ± 90 degrees. In this case, all modules' blocks align with a cubic lattice, all neighboring with opposite gender blocks (Fig. 1(b)). In addition, surface-to-surface contacts are maintained automatically, thereby ensuring easy connection control. On the other hand, a problem of collision among modules must be avoided through the appropriate design of a reconfiguration sequence.

2.2 Self-Reconfiguration

Distributed algorithms and representation of configuration were investigated in various studies of self-reconfiguration. Aside from general problems of total connection and collision, the M-TRAN system has its own difficulties because of its actuation degrees of freedom and surface-to-surface connections [12]. We conducted two approaches: design by humans and introduction of regularity in structure. As the first approach, several examples of reconfiguration were undertaken manually using the software tools described in Section 4.

The second approach is to introduce macro-scale regularity in structure. Similarly, with a "meta-module" idea, we designed several regular structures and motions (Fig. 2) [11]. A regular structure is assembled mostly using identical building blocks. Each building block can change its position while maintaining the total regularity after each predefined sequence of motions. As such, motions of building blocks can be made in parallel: parallel control is suitable. Therefore, one of our research objectives using M-TRAN is experimental verification of these motions using decentralized (autonomous) control.

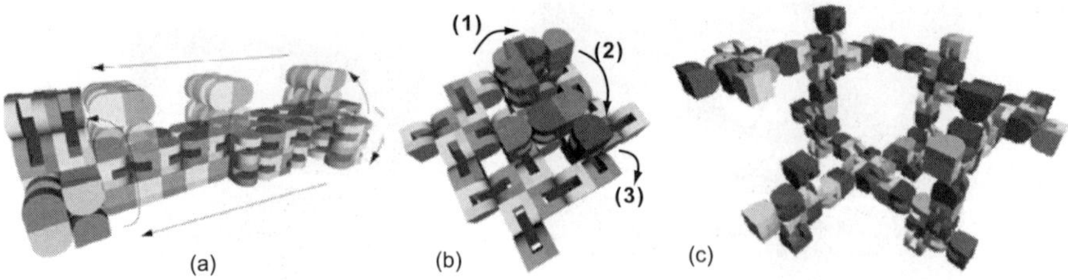

Fig. 2. One-dimensional to three-dimensional regular structures and motion sequences.

3 Hardware

We have developed 50 M-TRAN III modules (Fig. 1). Its size and weight (65 × 65 × 130 mm, 420 g) are similar to those of the former prototypes. The greatest improvement is in its faster and stronger connection mechanisms.

Instead of permanent magnets, which were used in former prototypes [12, 13], a new mechanical connector was designed based on the latch connector in [15]. Four hooks extend from the surface and hold another module at its passive surface. The active block has three such mechanisms, each driven by a DC motor. This mechanism is faster (about 5 s) and less power-consuming (1 W for 5 s) than the former (about 1 min and 4 W for 1 min, respectively).

A battery and a power supply circuit are in the passive block of the module to allow stand-alone operation. The link part contains two geared motors to rotate two blocks, potentiometers to measure those angles and a driver circuit including a microprocessor, which controls two motors using PID control.

The module has four microprocessors: one main CPU (SH-II; Renesas Technology Corp.) and three slaves (H8; Renesas Technology Corp.). Each slave CPU is placed in each of the three mechanical parts: passive and active blocks and a link. This design reduces mutual wiring among these parts. Most mechanical control and sensing processes are executed using slave CPUs under supervision of the main CPU.

The main CPUs of all the connected modules mutually communicate via a Controller Area Network (CAN) bus. This type of global communication bus is adopted because its implementation is easier than neighbor-to-neighbor communication. A network bus connection is maintained through electrodes on the connection surfaces (Fig. 1 (a)). In addition, a wireless modem based on Bluetooth technology is used to send commands from the host PC.

Aside from two electrodes for the CAN bus, two electrodes are used on each connection surface, by which neighboring modules exchange one-bit information. Ten pairs of infrared transmitters and detectors, used as proximity sensors, are placed on both blocks' surfaces. A three-axis acceleration sensor in the passive block measures the direction of gravity.

4 Controller and Programming Tools

The control of self-reconfiguration using the former prototype was distributed, but globally synchronous. Using M-TRAN III, we intend to try various controls from a single master and globally synchronous one to a parallel asynchronous one. In addition, we intend to construct a homogeneous system. Therefore, we designed an onboard controller by which various types of control can be realized within the same framework. In addition, we have developed various tools to design sequences for self-reconfiguration.

4.1 Distributed Controller by Virtual CPU

The basic idea of the controller is producing a virtual CPU with a communication channel. Two important functions are remote control of another module and shared memory, by which even one module can control everything of all the modules as a master.

The virtual CPU was designed as an 8-bit CPU, having 32 KB program memory, 128 byte data memory, peripheral devices (joint motors, connection mechanism and sensors), and a communication channel (Fig. 3). The command interpreter is its core, which executes incoming commands either from the program memory, from another module (via the network), or from the host PC (via the network or the Bluetooth modem), in almost the same format. Remote execution and shared memory are provided by this mechanism.

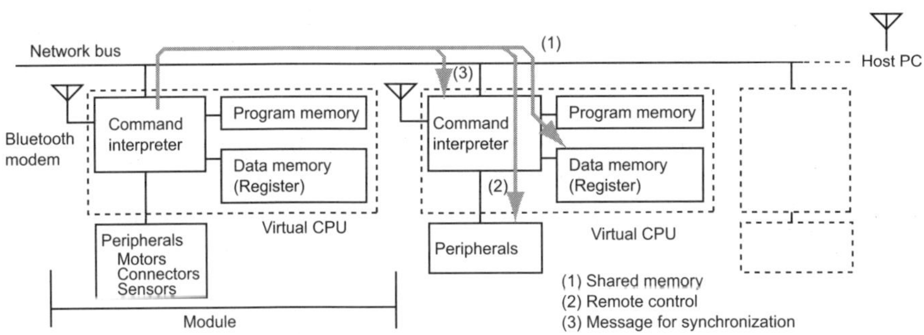

Fig. 3. Controller block diagram.

A self-reconfiguration sequence is stored as a program for this virtual CPU. Hereafter, such a program is called an SR program. An example of an SR program command is a three byte command, {'m',a,b}, in which the first byte, 'm', indicates rotation of two joints. The remaining two bytes are for angle values. A similar command, {'n',id,a,b}, is used for remote control. This command works similarly as the previous command if the second byte, the id, equals its own ID. Otherwise, the command interpreter sends a message (up to eight bytes) to the network as $\{sid, id, \text{'r'}, \text{'m'}, a, b\}$, in which the first and second bytes are ID numbers of the sender and destination modules (Fig. 3 (2)). The module specified by the second byte receives the message, interprets from the third byte, and executes the remote command ($\{\text{'m'}, a, b\}$), which achieves joint control. For motion synchronization, a special message is returned when such joint control is completed within a preset margin (Fig. 3 (3)). Shared memory is implemented using a similar mechanism (Fig. 3 (1)).

4.2 Programming and Emulation

An SR program is made in two ways: manual programming using command mnemonics and assembly using a programming tool, or automatic conversion using a kinematics simulator. The kinematics simulator previously developed [12, 14] is used to design a module configuration, design a sequence of reconfiguration, store the sequence in a script form (called an SR script hereafter), and verify it by kinematics and dynamics simulations. Conversion from an SR script to an SR program is made using an additional tool.

The format of an SR program differs from that of an SR script. Therefore, conversion is made as follows. While the kinematics simulator simulates motions according to the SR script, it records step-by-step changes in joint angles and connections in the form of an SR program. Consequently, the converted program is a series of remote controls that is globally synchronous and executed directly by a single master.

A program for decentralized and parallel control is made currently only manually. We have developed an emulator to verify the program before experimentation. The emulator is based on the kinematics simulator and emulates multi-virtual CPUs and network communications.

This emulator is also useful for development of the onboard program. Aside from kinematics simulation and network emulation, the virtual CPU program in the emulator is source-code compatible in C-language with the onboard one. Therefore, the same code is used in both the emulator and the onboard controller.

4.3 Initialization

Each module has a unique ID number by which one-to-one communication is made as described above. For all experiments, the same SR program was loaded into all modules from the host PC via the network bus. For each module to work appropriately, information regarding configuration is obtained using the startup program.

At startup, all modules start the same process, by which one module is selected as a master. Then under supervision by the master, each module identifies its neighbors using a one-bit channel described in Section 3, and the modules are counted. In addition, each module obtains its orientation using an acceleration sensor.

Through the exchange of neighbors' IDs and orientation, the total configuration and position and orientation of each module in the configuration is identified. A single master conducts this startup process, but some parallel processes exist for identification. After this process, the same master conducts all the controls, or parallel processes start in all the modules.

5 Experiments

Several experiments were carried out and performance of the system was verified. Experiments by a single master control include metamorphosis from a four legged walker to a linear form in Fig. 4, and repetitive self-reconfiguration of a linear structure in Fig. 5. Although a master module that was selected using the startup process conducted all the motions in these experiments, the same reconfiguration was realized by any master, even when the structure was assembled using different modules (different ID numbers) and in different local orientations.

Fig. 4. Metamorphosis from a four-legged walker to a linear form.

Fig. 5. Climbing a step using repetitive self-reconfiguration.

Figure 6 shows a simple experiment using decentralized and homogeneous control. Two modules execute the same SR program. They repeat joint control twice and connection and disconnection once for each cycle. Synchronization via message exchange is conducted. Thereby, connection (c) follows completion of joint control (b) and disconnection (d) follows connection of the other surface (c). The program is 33 command lines in mnemonic and 93 bytes.

Similarly, repetitive self-reconfiguration, like that shown in Fig. 7, was designed. A linear structure moves forward through repetition of local self-reconfiguration. The main idea of control is: (1) two modules, b and d in Fig. 7 for example, move cooperatively; (2) f and h start motion by a message from the former modules; (3) four modules (a, c, e, g) on the other side wait for completion of the pairwise motions to keep total connection; then, (4)

Fig. 6. Parallel control.

after completion, modules on the left and right sides change their roles locally and begin the process from (1). Although all modules are identical, there are initially only two modules (Fig. 7 a, h) which have the fewest neighbors. Those two are distinguishable by their block type (color in Fig. 7) on the end. The direction of the motion is therefore determined. They initiate the process after the startup process.

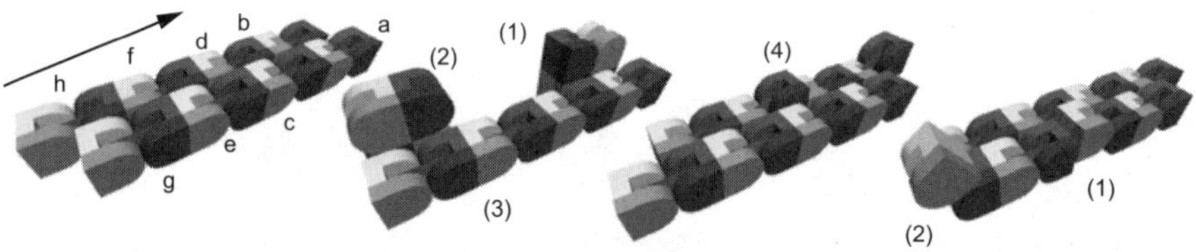

Fig. 7. Locally synchronous repetitive metamorphosis.

Experiments were carried out using from 4 to 20 modules (Fig. 8), all by the same program (231 command lines in mnemonic and 610 bytes). In each cycle, each module controls connection 10 times and the structure moves two module length. The controller in each module uses only local information (four neighbors' IDs) and local synchronization with neighbors.

Another decentralized self-reconfiguration similar to Fig. 2 (b) was tried. Two modules (a,b in Fig. 9 (1)) on the edge ascend, move on the plane made of other modules (Fig. 9 (2), (3)), and reach the other edge ((4)). Then other two (c,d) follows ((5)–(7)). The SR program is 394 command lines in mnemonic and 963 bytes.

There still remain problems. Some trials of experiments in Fig. 8 (d) and Fig. 9 failed either by a communication failure or by misalignment caused by gravitational force. The latter problem is not simple, as alignment accuracy depends on mechanical rigidity, joint control performance, and sequences of reconfiguration. However, most of the problems will be solved by improvement of software and sequence design.

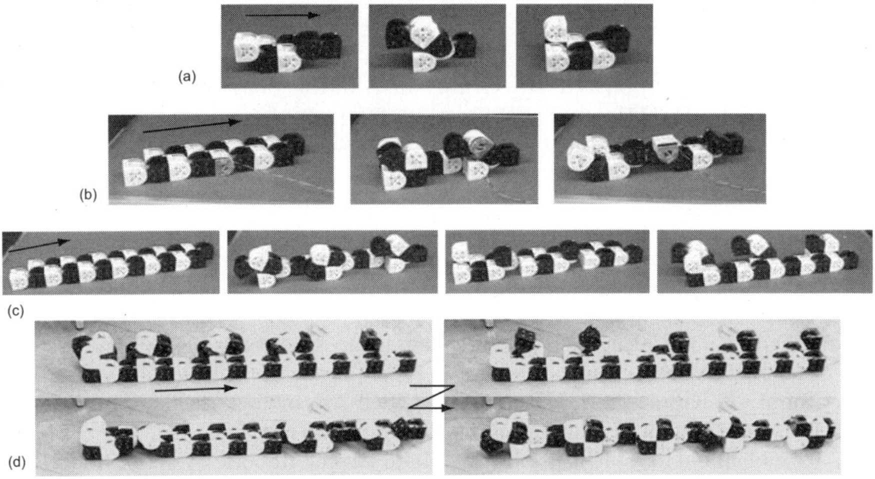

Fig. 8. Experiments of distributed metamorphosis (linear).

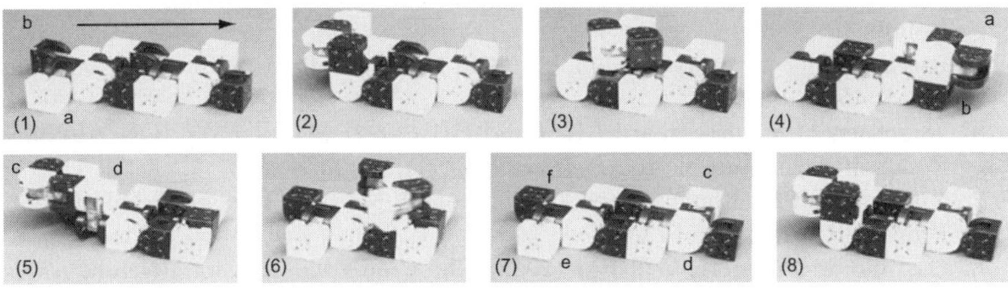

Fig. 9. Experiment of distributed metamorphosis (planar).

6 Conclusion

We developed a new M-TRAN system, an integrated distributed controller, and programming tools. Experiments were carried out, which include self-reconfiguration of up to 20 modules requiring more than 50 motion steps and 100 disconnections and reconnections. In all experiments, homogeneity of the total system was maintained: all modules were the same in programs and data. In addition to experiments by centralized and globally synchronous control, decentralized and locally synchronous control was tested. The results verified the performance of our modular robot M-TRAN and the control software.

Currently, the reconfiguration sequence is designed manually. For future research, we will explore various means of automatic programming. Autonomous reconfiguration using sensor information is another subject. Though the structures in Fig. 8 and in Fig. 9 only move straight in the experiments, they have the ability to change its direction as in Fig. 5 and Fig. 2 (b). Using proximity sensors, we will attempt to let the modules determine their direction of motion individually.

Acknowledgements

This study was supported by Grants-In-Aid for Scientific Research (No. 16300063) from the Ministry of Education, Culture, Sports, Science and Technology (MEXT) of Japan. We thank Mr. T. Hasuo of Tsukuba University for development of hardware and controller programs.

References

1. Fukuda T and Nakagawa S (1988) Dynamically Reconfigurable Robotic System, Proc. ICRA'88.
2. Murata S, Kurokawa H, Kokaji S (1994) Self-Assembling Machine, Proc. IEEE Intl. Conf. Robotics and Automation (ICRA), pp 441-448.
3. Chirikjian G (1994) Kinematics of a Metamorphic Robotic System, Proc. IEEE Intl. Conf. Robotics and Automation (ICRA), pp 449-455.
4. Ünsal C, Kiliççöte H, Khosla PK (2001) A Modular Self-Reconfigurable Bipartite Robotic System: Implementation and Motion Planning, Autonomous Robots 10:23-40.
5. Rus D, Butler Z, Kotay K, Vona M (2002) Self-Reconfiguring Robots, Commun. ACM, 45-3:39-45.
6. Jørgensen MW, Østergaard EH, Lund HH (2004) Modular ATRON: Modules for a self-reconfigurable robot, Proc. IROS04, pp 2068-2073.
7. Yim M, Duff D, Roufas K (2000) Polybot: A Modular Reconfigurable Robot, Proc. ICRA 2000, pp 514-520.
8. Castano A, Behar A, Will PM (2002) The Conro Modules for Reconfigurable Robots, IEEE/ASME Trans. Mech. 7-4:403-409.
9. Butler Z, Kotay K, Rus D, Tomita K (2004) Generic Decentralized Control for Lattice-Based Self-Reconfigurable Robots, Intl. J. Robotics Research, 23-9:919-937.
10. Østergaard EH, Lund HH (2004) Distributed Cluster Walk for the ATRON Self-Reconfigurable Robot, IAS-8 pp 291-298.
11. Østergaard EH, Tomita K, Kurokawa H (2004) Distributed Metamorphosis of Regular M-TRAN Structures, Proc. DARS2004, pp 161-170.
12. Murata S, Yoshida E, Kamimura A, Kurokawa H, Tomita K, Kokaji S (2002) M-TRAN: Self-Reconfigurable Modular Robotic System, IEEE/ASME Trans. Mech. 7-4:431-441.
13. Kurokawa H, Kamimura A, Yoshida E, Tomita K, Murata S, Kokaji S (2003) M-TRAN II: Metamorphosis from a Four-Legged Walker to a Caterpillar, Proc. IROS03, pp 2452-2459.
14. Kamimura A, Kurokawa H, Yoshida E, Murata S, Tomita K, Kokaji S (2005) Automatic Locomotion Design and Experiments for a Modular Robotic System, IEEE/ASME Trans. Mech, 10-3:314-325.
15. Terada Y, Murata S (2004) Automatic Assembly System for a Large-Scale Modular Structure, Proc. IROS04, pp 2349-2355.

Preliminary Results in Tracking Mobile Targets Using Range Sensors from Multiple Robots

Elizabeth Liao, Geoffrey Hollinger, Joseph Djugash, and Sanjiv Singh

Carnegie Mellon University {eliao@andrew.cmu.edu, gholling@andrew.cmu.edu, robojoe@cmu.edu, ssingh@ri.cmu.edu}

Summary. In urban search and rescue scenarios, human first responders risk their lives as they routinely encounter hazardous environments. A team of robots, equipped with various sensors, deployed in such an environment can be used to track emergency personnel such as firefighters, reducing the risk to human life. This paper explores techniques for tracking a mobile target and coordinating a team of robots, equipped with range-only sensors, through smoke-filled, high-temperature environments. The particular strengths of our tracking and cooperative control algorithms are identified through a set of simulated examples.

1 Introduction

Smoke, darkness, and clutter handicap first responders during time critical emergency situations. Robot teams can help by providing feedback on environmental hazards and tracking human agents in the environment. Creating and maintaining a communications network and tracking the location of emergency response personnel are examples of some of the tasks that the robot teams can perform.

Tracking a moving target using a single robot in a near-ideal, clutter-free environment is a solved problem. Environments with dense obstacle configurations, however, can enforce constraints that limit the robot's tracking capabilities. Since speed and efficiency are critical, the use of multiple robots also becomes desirable. Although the applications for a robot team's assistance are numerous, a necessary functionality of a robot team in this scenario is the ability to reliably and accurately track a human over time. In order to achieve this higher tracking accuracy, the proper coordination of the robot team is important and mandatory. In this paper we explore these two separate yet inter-related problems, tracking a mobile target and coordinating a robot team to further improve the achieved tracking accuracy.

At a high level, the job of a robot team is to track a human while he/she is within range of the robot team. In order to perform such a task, the robot team requires sensors that can provide sufficient information to help localize the human. Cameras, sonar and other commonly used sensors fail to provide the necessary robustness and accuracy in the presence of smoke, darkness, and debris. Radio then becomes the obvious choice for our particular problem. Radio based sensors provide a unique advantage over the more commonly used sensors by delivering reliable measurements even under the presence of heavy smoke and debris. For these reasons, the results in this paper focus on the use of radio ranging sensors for tracking mobile targets.

The tracking and coordination algorithms presented in this paper are backed by a set of experiments conducted in simulation. Experiments in team coordination are performed with a four member robot team. Given only a floor map and an estimate of the robot positions, the robot team achieves reliable tracking of the mobile target.

2 Related Work

Robotics researchers have examined the potential for multiple robot assistance in urban search and rescue environments [9], and previous research has explored the use of multiple robots in indoor environments for mapping and intruder detection [6].

Recent advances in radio technology allow for range-only measurements between inexpensive transponders and receivers. These sensors can provide information in dynamic, noisy environments without the necessity of line-of-sight. To take advantage of these sensors, algorithms must be developed for range-only localization and tracking. Kantor and Singh present preliminary work in localizing robots using range-only measurements using an extended Kalman filter (EKF) and a particle filter in [7]. Djugash et al. give further research and testing of range-only localization and mapping in [2].

Other researchers have examined the potential for using prior knowledge from the map during tracking and localization. Liao et al. presents a system for limiting particle paths to the voronoi diagram of the map space in [12]. Additionally, Min Oh et al. have described a system for tracking humans in outdoor environments using GPS and map-based priors [14]. These papers use sensors that provide both range and bearing information, and they explore map environments with limited complexity.

Tracking using formation control of a multi-robot team can be adapted for the human tracking scenario. The system described by Saber et al. [15] tracks in formation by generating cost functions based on formation control, collision avoidance, and tracking. Fierro et al. present a framework for the cooperative control of a robot that is broken up into two major parts: control law selection and trajectory generation [3]. Naffin and Sukhatme present a method to dynamically grow and maintain formations through negotiations [13]. While these approaches address maintaining formations in free space with relatively few obstacles, they do not explicitly mention how they handle cluttered environments.

3 Tracking with Range-Only Data

The task of tracking humans with range-only data introduces severe nonlinearities into any tracking algorithm. In addition, the lack of human odometry and motion constraints further complicate the problem. This section presents two algorithms for tracking humans using range-only data. The first is a simple particle filter, while the second incorporates the knowledge of a map to improve the accuracy of the tracker.

3.1 Particle Filter Tracking

One technique for tracking using range-only sensor data can be derived using Monte Carlo methods. Our algorithm (a modified version of the one presented for localization by Kantor and Singh [7]) uses many particle estimates of the first responder's position, propagates these particles randomly, and then moves them towards the high probability regions provided by the range measurements. This calculates the maximum a posteriori estimate of the first responder's position.

Letting $x_p(k)$, $p \in \{1, 2, \ldots, N_p\}$, be a distribution of particles at time k, where N_p is the number of particles in that distribution, the algorithm for updating the particle filter proceeds as follows:

1. Propagate the state of each particle by computing $\widetilde{x}_p(k) = x_p(k) + \omega_p(k)$ where ω_p is generated as zero-mean Gaussian noise with covariance R.
2. Assuming that each measurement is independent, compute the weight of each particle from the product of the Gaussian PDFs associated with current range measurements $(m_r(k))$ from each robot
$$w_p = \prod_r p(r_p|m_r(k)) \tag{1}$$
where r_p is the distance between $x_r(k)$ and $\widetilde{x}_p(k)$.
3. Normalize the weights so that $\sum_{p=1}^{N_p} w_p = 1$.
4. Resample the particles such that each $x_p(k+1)$ particle is selected from $\widetilde{x}_i(k)$, and the probability of selecting each particle is w_i. The resulting distribution moves towards areas of high probability based on measurements from all robots.

The particle filter provides a method for tracking humans without the use of odometry data. Additionally, it avoids any linearization of the range measurement probability distribution function.

3.2 Using Prior Knowledge from the Map

To fully utilize the benefits of operating in an indoor environment, we extend our particle filter implementation to incorporate prior knowledge of the map. Given an estimate of the firefighter's position, the uncertainty of subsequent measurements is bounded by obstacles on the map. For instance, if we know that a firefighter is traveling in a hallway, subsequent estimates should remain in that hallway. To take advantage of this insight, we make two modifications to the particle filter.

1. Particles are restricted from moving through walls. When calculating particle weights in Step 2, particles that have moved through walls are discounted by ten orders of magnitude.
2. Under the assumption that firefighters generally move in the center of corridors, the weights of particles are adjusted such that they equalize their distance to walls. This is accomplished by viewing the map as an additional 2D Gaussian measurement. The mean of this Gaussian is given by the center of the current corridor, and the standard deviation is determined by the square of the width and length of the corridor. The use of a Gaussian provides a principled method for integrating the information from the map into the particle filter while still maintaining the assumption that firefighters tend to remain in the center of hallways. The use of a uniform probability distribution function, for instance, would not take this assumption into account. This changes the weight update step (Step 2) to the following:

$$w_p = p(\widetilde{x}_p(k)|M) \prod_r p(r_p|m_r(k)) \tag{2}$$

where $p(\widetilde{x}_p(k)|M)$ is the probability of the particle's position given the Gaussian measurement from the map.

4 Multi-Robot Coordination

One method for increasing the accuracy of firefighter tracking with range-only sensors is through cooperative control of a multi-robot team. The goal of a cooperative control algorithm is to assist in tracking by producing desirable team configurations to reduce the uncertainty of the firefighter's position.

Before trying to position robots into configurations, a metric is needed to identify *good* configurations. Preferable configurations will magnify sensor error less, assuming that ranging devices are inherently noisy. We use a metric called the Geometric Dilution of Precision (GDOP), which relates the configuration of ranging devices to location error. Kelly uses Dilution of Precision by evaluating a robot's position error using landmark configuration for localization purposes [8]. We are in essence using a similar technique in reverse by having robots be the "landmarks". A two dimensional version of the GDOP formulation [1] was used such that only longitude and latitude were incorporated into the calculations. Since the problem domain is only in two dimensions, we refer to 2D GDOP as GDOP for the remainder of this paper.

Fig. 1 shows example robot team configurations and the corresponding GDOP values. The firefighter is located in the center while the angles between the four team members are varied to form different configurations.

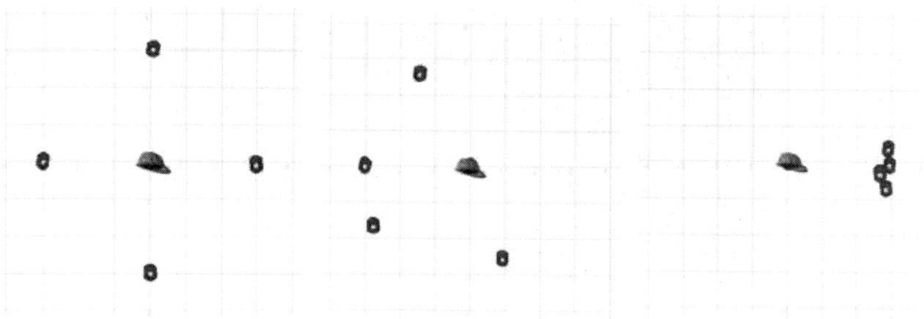

(a) GDOP=1.000 (b) GDOP=1.134 (c) GDOP=3.724

Fig. 1. Example four-robot configurations and their corresponding GDOP values. The values range from the best(left-most) to least desirable(right-most)

4.1 Cooperative Control Algorithm

In the cooperative control algorithm, each robot considers its teammates as part of the environment (stationary) and creates a plan to position itself with that belief. Here, a *plan* is defined as a path consisting of a series of waypoints ending with a desired goal. The only information available to create the plan are the positions of other team members and the firefighter. Key decisions of the algorithm are summarized below and are detailed by Liao [11].

4.2 Cost Map

A discretized cost map is used to decouple global planning from local navigation. Once a plan is produced, it can be passed off to another algorithm for local obstacle avoidance and waypoint following. Each robot initializes its cost map using an occupancy grid representing the floor plan. Every grid cell represents a possible goal location and is assigned a cost based on configuration of the team and distance from the firefighter assuming that the robot was located in that cell. The most desirable location of the robot corresponds to the cell with the lowest cost, while the best path to the goal is generated using a wavefront propagation path planner [10].

A weighted distance from the robots to the firefighter needs to be incorporated with the configuration cost into an overall cost so that robots are not too close to obstruct the human's movement but not so far so that they are out of sensor range. This weighted wavefront distance function assigns the lowest value to an ideal distance and then increases in value as robots get farther from the ideal distance.

For the configuration policy, we developed the *Mean Angle* policy, which produces a score according to how much the configuration deviates from an equal angle spacing between robots. Mean angle can be calculated using the following equation:

$$\sum_{i=1}^{n} \exp \left| \frac{360^o}{n} - \angle i, mod(i+1, n) \right| - 1$$

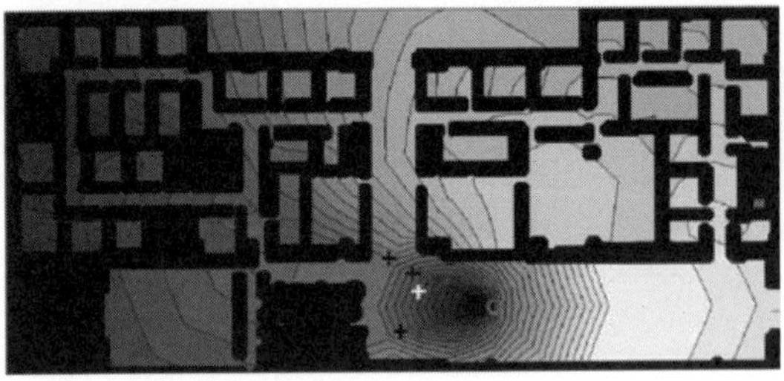

Fig. 2. Cost Contour map using weighted wavefront and mean angle policies. The light regions are the most favorable for the robot represented by the white cross.

This equation calculates a score by looking at the angles between adjacent robots with the firefighter as the vertex. Larger differences between an angle and the ideal angle will contribute to a higher score.

After integrating the Mean angle configuration cost with the weighted wavefront distance, the overall cost map is produced. Fig. 2 shows an overall cost contour map from the perspective of the white robot (far right cross). The light regions symbolize lower (better) cost, and dark regions symbolize higher (worse) cost. The black crosses represent other members of the robot team, and the circle represents the firefighter.

5 Simulation Results

Our cooperative control algorithm was implemented using software from the Player/Stage Project, an open source project that supports research in robots and sensors [4, 5]. Each robot executes the cooperative control algorithm by having an individualized version of the cost map. Initially the cost map is populated with obstacles contained in the provided floor plan. The distance and angle based costs are then filled into the remainder of the grid cells.

At every planning iteration, after obtaining the new positions of team members and the firefighter, the robot updates its cost map. The wavefront planner then generates a new path, which is passed off to the built-in Vector Field Histogram (VFH) [17] driver in Player/Stage for waypoint following and local obstacle avoidance. Since the algorithm implementation relies on frequent replanning, robots almost never reach their current goal and will typically choose different goals at every iteration.

The runtime of this implementation is highly dependent on how fine the floor plan is discretized in the cost map. The size of the cost map is an important factor to consider because every cell in the cost map is updated at the beginning of every iteration. Then, an exhaustive search over the cost map is used to locate the best position for a robot. Finally, the cost map is used again with wavefront planner to find the lowest cost path to the lowest cost cell.

This cooperative control algorithm is similar to Stroupe's Move Value Estimation for Robot Teams (MVERT) architecture for action selection in multi-robot teams [16]. While a robot running MVERT tries to predict the actions of its teammates to plan for the next step, our cooperative control algorithm treats teammates as stationary objects and makes an entire plan to the goal.

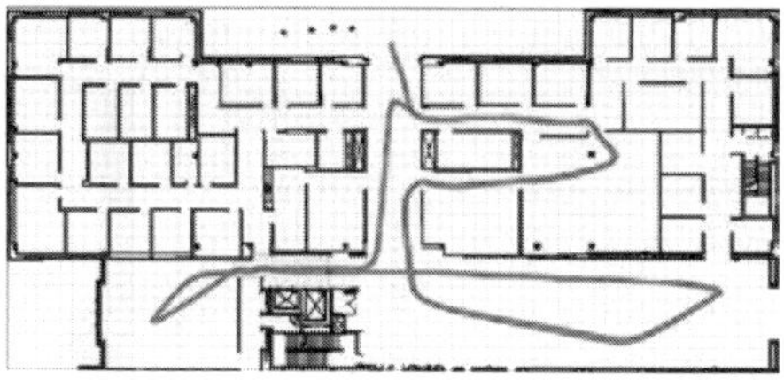

Fig. 3. Floor plan with initial robot positions and firefighter path. Initial position of the firefighter is the topmost point of the path.

5.1 Experiments in Player/Stage

Currently, many real-world applications use sonar based line-of-sight (LOS) range sensors, but we would also like to investigate the added benefits of sensors that can range through walls (non-LOS ability). To quantify the difference between LOS and non-LOS ranging performance, experiments were performed using GDOP as the evaluation metric. Only robots with LOS to the firefighter contributed to GDOP in the appropriate experiments.

For each experiment, a robot representing the firefighter traveled on a predetermined path while a team of four robots continually positioned themselves according to the commands generated by the cooperative control algorithm. During each iteration, GDOP was calculated using only the firefighter position and robot positions within sensor range of the firefighter. Since the algorithm ideally performs better without obstacles, experiments were run comparing the original floor plan (Fig. 3) to ones run using an "empty" floor plan (the original floor plan without interior walls).

Histograms of the experiments (Fig. 4) show bins of GDOP values recorded during the entire firefighter path. The x-axis shows GDOP bin values while the y-axis shows the normalized histogram counts. Normalization was needed to compare experiments with different length data sets. The percent valid shown on each graph is the amount of time a valid GDOP value was obtained(i.e. when 2 or more robots are in range). Although histograms 4(a) and 4(b) should intuitively be identical, the discrepancy arises by having outside building walls remain in the "empty" floor plan. In the LOS case on an empty floor plan, it is possible that the exterior building walls block LOS to firefighter. This may in turn lead to configurations that evaluate to higher GDOP values.

5.2 Simulated Tracking Results

We ran simulated tests at various noise levels to determine the performance of the particle filter tracking with and without the map. Table 1 gives the characteristics of the three noise levels. During the runs, a simulated first responder moves throughout the entire map including both open and cluttered spaces (a total space of about 50x30 meters). Four robots move around the map using the coordination strategy described in Sect. 4 and attempt to localize the firefighter using simulated ranging measurements. Since the measurements are meant to model ranging radio, line-of-sight is not taken into account. The positions of the robots are assumed to be known for these tests. We set the particle filter to use 100 particles, which provides a reasonable loop speed.

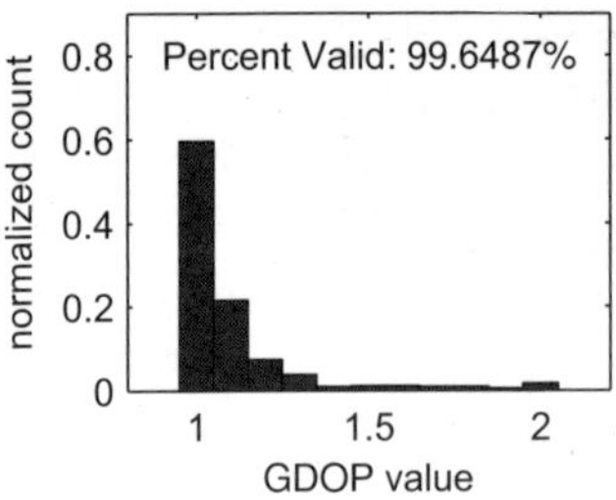

(a) Line-of-sight ranging on an "empty" floorplan

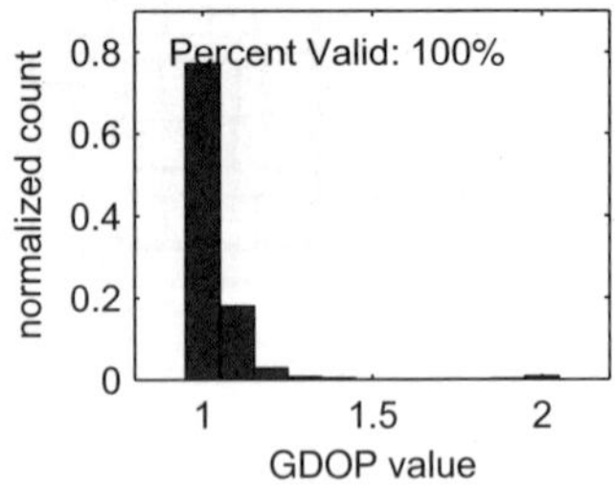

(b) non Line-of-sight ranging on an "empty" floorplan

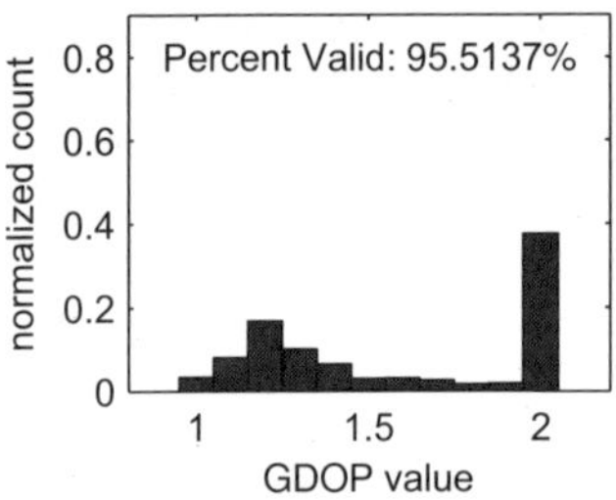

(c) Line-of-sight ranging on the original floorplan

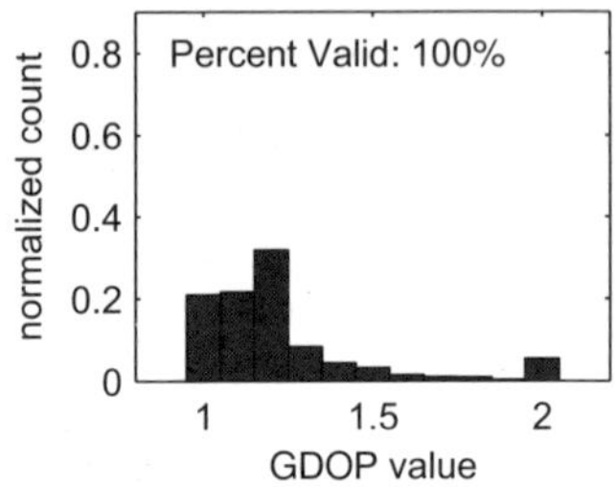

(d) non Line-of-sight ranging on the original floorplan

Fig. 4. Histogram representing GDOP values of a single experiment run. The x-axis shows the GDOP bin values and y-axis shows the normalized bin counts.

Table 1. Characteristics of three simulated noise levels and the average Euclidean errors for various simulated tracking algorithms

	Simulated sensor variance (m^2)	Dropped meas. (%)	Avg PF error (m)	Avg map PF error (m)
Low noise	0.5	25	0.2429	0.2295
Moderate noise	1.0	50	0.4163	0.3982
High noise	1.5	75	0.7461	0.6720

In addition, Table 1 also gives the average Euclidean error over the entire run (approximately 10,000 estimates) for all runs. Using prior knowledge from the map improved the results, and this improvement was more pronounced at higher noise levels. Fig. 5 gives sample histograms of the Euclidean error for the tracking runs.

Additional tests (not presented) were performed using a Pioneer robot and sonar sensors to verify the range-only tracking algorithms on actual hardware. These experiments confirmed the simulated results.

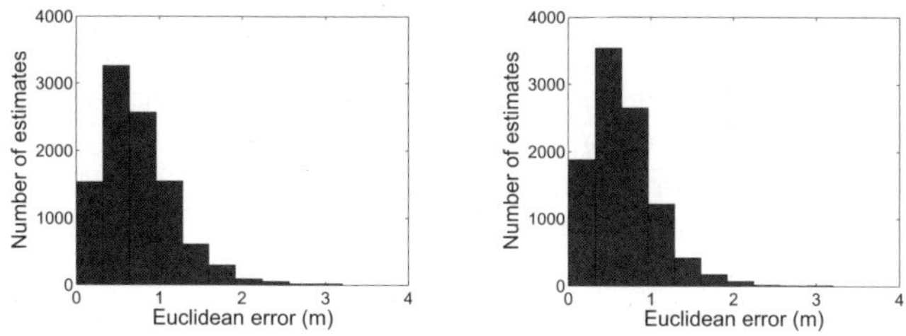

Fig. 5. Sample histograms of Euclidean error at high noise level for simulated tracking using particle filter without map (left), and particle filter augmented with the map (right)

6 Conclusion/Future Work

Our results show that a particle filter provides feasible tracking using noisy, range-only sensor data. The filter is robust enough to avoid divergence due to high sensor variances, dropped measurements, and sensor silences. It has been revealed that using prior knowledge from the map has the potential to further increase the tracking accuracy of the particle filter. However, if robots are operating in scenarios in which the map is not known (or only partially known), the particle filter without the map provides acceptable tracking performance.

To improve tracking accuracy, a cooperative control algorithm was developed for a robot team to minimize the position uncertainty of the target. Using a decentralized cost map approach, the cooperative control algorithm locates the lowest cost position for the robot and then finds the lowest cost path to reach that position. Configurations were evaluated using the GDOP metric, a measure of how team configuration will magnify sensor error. Using experiments running the cooperative control algorithm in the Player/Stage simulation environment, the effect of LOS vs. non-LOS sensors on team performance were compared. In a cluttered environment, the use of non-LOS sensors improved performance by giving the robot team the ability to form preferred configurations.

For future research, we plan to examine alternative methods for utilizing prior knowledge from the map. Algorithms like Hidden Markov Models have the potential to more accurately model the tendencies of human first responders in environments with maps. Such an approach would also allow robots to learn motion patterns from human movement data.

To deal with real scenarios, robot failure detection could improve the cooperative control algorithm's robustness and flexibility. Robot failures are important to detect because in hazardous environments, malfunctions are common. For robustness, a team should be able to detect if a robot has stopped or is out of range and compensate for the loss in its behavior.

Acknowledgements

We thank Bradley Hamner for sharing his hardware and software optimization expertise.

References

1. P. Dana. Global Positioning System Overview, The Geographer's Craft Project, Department of Geography, The University of Colorado at Boulder, http://www.colorado.edu/geography/gcraft/notes/gps/gps.html.
2. J. Djugash, S. Singh, and G. Kantor. Range-only slam for robots operating cooperatively with sensor networks. In *Proceedings, IEEE International Conference on Robotics and Automation*, 2006.
3. R. Fierro, P. Song, A. Das, and V. Kumar. *Cooperative Control and Optimization*, volume 66, chapter Cooperative control of robot formations, pages 73–93. Kluwer Academic Press, 2002.
4. B. Gerkey, R. Vaughan, and A. Howard. The Player/Stage project: Tools for multi-robot and distributed sensor systems. In *Proceedings of the International Conference on Advanced Robotics (ICAR 2003), Coimbra, Portugal, June 30 - July 3, 2003*, pages 317–323, 2003.
5. B. Gerkey, R. Vaughan, A. Howard, and N.Koenig. The Player/Stage Project, http://playerstage.sourceforge.net/.
6. A. Howard, L. Parker, and G. Sukatme. Experiments with a large heterogeneous mobile robot team. *Intl. Journal of Robotics Research.*, 2005.
7. G. Kantor and S. Singh. Preliminary results in range-only localization and mapping. In *Proc. of IEEE Conf. on Robotics and Automation.*, 2002.
8. A. Kelly. Precision dilution in triangulation based mobile robot position estimation. In *Intelligent Autonomous Systems*, 2003.
9. V Kumar, D. Rus, and S. Singh. Robot and sensor networks for first responders. *Pervasive Computing*, pages 24–33, 2004.
10. S. M. LaValle. *Planning Algorithms.* Cambridge University Press (also available at http://msl.cs.uiuc.edu/planning/). To be published in 2006.
11. E. Liao. Cooperative control of a multi-robot team in a firefighting scenario. Technical Report CMU-RI-TR-05-50, Robotics Institute, Carnegie Mellon University, Pittsburgh, PA, in progress.
12. L. Liao, D. Fox, J. Hightower, H. Kautz, and D. Schulz. Voronoi tracking: Location estimation using sparse and noisy sensor data. In *Proc. Intl. Conf. on Intelligent Robots and Systems*, 2003.
13. D. Naffin and G. Sukhatme. Negotiated formations. In *Proceedings of the Eighth Conference on Intelligent Autonomous Systems*, pages 181–190, 2004.
14. S Min Oh, S. Tariq, S. Walker, and F. Dellaert. Map-based priors for localization. In *Proc. Intl. Conf. on Intelligent Robots and Systems.*, 2004.
15. R. Saber, W. Dunbar, and R. Murray. Cooperative control of multi-vehicle systems using cost graphs and optimization. In *Proc of the American Control Conference, June 2003.*, 2003.
16. A. Stroupe. *Collaborative Execution of Exploration and Tracking Using Move Value Estimation for Robot Teams (MVERT).* PhD thesis, Robotics Institute, Carnegie Mellon University, Pittsburgh, PA, September 2003.
17. I. Ulrich and J.Borenstein. VFH+: Reliable obstacle avoidance for fast mobile robots. In *Proceedings of the 1998 IEEE International Conference on Robotics and Automation.*, pages 1572–1577, 1998.

Robotic Swarm Dispersion Using Wireless Intensity Signals

Luke Ludwig[1,2] and Maria Gini[1]

[1] Dept of Computer Science and Engineering, University of Minnesota
(ludwig,gini)@cs.umn.edu

[2] BAESystems Fridley, MN

Summary. Dispersing swarms of robots to cover an unknown, potentially hostile area is useful to setup a sensor network for surveillance. Previous research assumes relative locations (distance and bearing) of neighboring robots are available to each robot through sensors. Many robots are too small to carry sensors capable of providing this information. We use wireless signal intensity as a rough approximation of distance to assist a large swarm of small robots in dispersion. Simulation experiments indicate that a swarm can effectively disperse through the use of wireless signal intensities without knowing the relative locations of neighboring robots.

1 Introduction

Deploying large numbers of small simple robots in a decentralized swarm-like fashion is gaining recognition and popularity in many problem domains. One of the primary issues is how the swarm of robots will move. A common task is to spread out and cover an unknown area as thoroughly and quickly as possible in order to setup a sensor network for surveillance. This may be useful in areas hostile to humans such as disaster or military zones, or even planetary exploration.

In a swarm approach each robot is small, simple, and executes the same software program. Swarm methods bring many advantages. Since they are decentralized there is no single point of failure that can bring the entire system down. In fact many of the robots can fail and the swarm system will still function. Due to the simplicity of the robots they will be cheap to manufacture, another reason they are expendable. The designer of a swarm hopes that through individual actions based on local decisions the swarm as a whole will produce the intended emergent behavior. As in the natural world, unexpected emergent behavior may even occur.

A common theme in robot dispersion research is a connection to behavior in the natural world, whether it be of a biological, chemical, social, or physics based behavior. In particular there are many approaches that draw on the general concept of repulsive and attractive forces between robots [13, 2, 9]. There are subtle differences between these approaches, but the general concept is the same; move away from neighboring robots, but not too far away. All of these approaches assume that the relative locations (distance and bearing) of neighboring robots is available to the robot through its sensors. This can be obtained using a 360° laser range finder or an omnidirectional camera. Size is the limiting factor. The robots must be large enough to carry the laser range finder or have enough processing power to analyze the images from the camera which is inherently processor intensive. A common laser range finder, the SICK LMS 200, has dimensions of 15 cm x 18 cm x 15 cm. This is a rather large payload for a robot such as the University of Minnesota's Scout, which is only 11 cm long by 4 cm wide [12]. Small robots have the advantage of being cheaper, simpler, and less noticeable. Bob Grabowski from Carnegie Mellon has classified existing small robots by size in his Small Robot Survey, and at least half of the robots in this survey would be incapable of sporting either a 360° laser range finder or an omni camera [6].

The question we attempt to answer is whether or not a swarm of small robots can be dispersed effectively without knowing the relative locations of neighboring robots. In particular, we use wireless signal intensity as an approximation of distance to assist in the dispersion. This requires a wireless 802.11 card on the robots, which is considerably smaller than the 360° laser range finder.

Theoretically, signal intensity varies according to the law of inverse signal propagation, which simply means the signal intensity is proportional to the inverse square of the distance it travels. In a practical setting, the environment plays a huge role by providing obstacles that cause noise in this signal. However, it is unnecessary for the signal intensity to be very accurate in order to provide some indication to a robot for which way it should travel. The primary property needed is for the signal intensity to decrease over time as the distance between robots increases, and vice-versa.

2 Related Work

In 1992 Gage was the first to consider the problem of area coverage by a team of robots [4]. He categorized the problem into three types: blanket coverage, barrier coverage, and sweep coverage. Blanket coverage, the most similar to our research, has the objective of maximizing the total area covered by a static arrangement.

A promising experiment by Howard, Mataric, and Sukhatme considers how to deploy a mobile sensor network in an unknown hostile environment [9]. They use robots equipped with a 360° laser range finder (4 meter range). No wireless communication is done. The dispersion is based on potential fields. Basically robots are repulsed by other robots and walls. They showed how a mobile sensor network can be deployed without centralized control, localization, or communication, using only local rules based on potential fields. Their results are impressive, but this approach is not possible for very small robots due to the large sensors required.

Hsiang et. al [10] use a leader-follower approach based on local rules where the robots form chains emanating from a single source of robots. The robots attempt to follow walls by keeping the walls on their left. Their simulation experiment was ran in a discrete grid world and assumes "local sensors." It would be interesting to see if this algorithm could operate well in a more realistic simulation environment, such as provided by Player/Stage [5], while using only small proximity sensors (ex infrared) for following robots.

Batalin and Sukhatme [1] rely on the deployment of beacons into the environment to help coordinate a decentralized algorithm that uses only local interactions between the robots and beacons to cover an unknown area. To use this approach robots must be large enough and capable of carrying the static beacons. It is more flexible to mobilize and miniaturize everything. Any small robot can accomplish the same task as a static beacon by simply remaining stationary. One of the key aspects of the algorithm presented in this paper is deciding when a robot in motion should stop and become a stationary beacon.

Research has been done on utilizing wireless signal intensity to approximate distance for localization of sensor networks. Haeberlen et. al [7] show how accurate localization at the room-level can be obtained in an office setting in which signal intensity data has been predetermined. Tian He et. al. [8] examines the limitations of range based localization and proposes a novel range-free localization scheme.

3 Clique-Intensity Algorithm

We assume the robots have a few small and simple proximity sensors that extend at least a meter (ex. infrared) that allow the robot to avoid most collisions with walls and other robots. We also assume that the robots have a wireless 802.11 card and are capable of obtaining signal intensity measurements with incoming packets. This is a standard requirement of the 802.11 interface. No other sensors are needed during dispersion, although most likely robots will be carrying some form of a camera or other sensor which is meant to be utilized once the sensor network is in place. The processing power on small robots is limited, which makes analyzing images from a camera during real-time motion difficult. For many applications, it is likely the camera exists solely as a means of communicating images back to some central node for further processing or even human-in-the-loop analysis.

There are many ways to use wireless signal intensity to aid a swarm in dispersing throughout an unknown environment. In comparison to all of the repulsive/attractive dispersion research in which relative distance and bearing of neighboring robots is known [13, 2, 9], signal intensity gives only a rough approximation of distance and no bearing information. This signal intensity must be tracked over time to determine which direction the robot should move. In a swarm of robots, each one may be in contact with many neighbors at a time. If it is known that one of the neighbors was stationary, then a robot could specifically reference the stationary robot's intensity and attempt to move in a direction of decreasing signal intensity until some threshold is reached. This is a key concept in the algorithm we developed.

The Clique Intensity Algorithm[3] is designed for a distributed homogeneous swarm, therefore the algorithm operates and runs from the perspective of a single robot in the swarm. The knowledge of each robot is a graph with robots as nodes and signal intensities between robots as weights. This graph is referred to as the connectivity graph. Robots share portions of their connectivity graphs with their neighbors such that each robot has the knowledge it needs to execute the algorithm. A clique is a graph or subgraph in which every node is connected to every other node. A maximal clique is not a subgraph of another clique. For each maximal clique in the connectivity graph a single robot is chosen to be the sentry for the clique, meaning it remains stationary. The other robots in the clique attempt to move away from the sentry, which is done by monitoring the change in the signal intensity over time. Each robot behaves in such a way that causes the entire swarm to disperse in an attempt to create cliques in the connectivity graph of size three or two. This is an attempt to triangulate the map which is known to be the most effective static configuration for the area coverage problem [11]. The algorithm is roughly composed of five basic steps (Figure 1).

Each robot in motion needs a sentry from which it monitors the signal intensity over time to determine which way to move. The primary decision to make is whether or not the robot is a sentry, and if not then the robot must decide which neighbor will be its sentry. This decision is made individually by each robot examining its connectivity graph and following a set of rules. The rules are structured such that each robot will arrive at the same decision as to which are sentries and which are in motion. Communication between robots of a bartering nature could be used to resolve the decision of which ones are sentries. This was not done since an attempt was made to avoid communication overhead and to keep the algorithm as simple as possible. The only communication between robots is the sharing of knowledge described above. See Figure 2 for a detailed description of the algorithm.

Considering the five steps in Figure 1, steps 1 and 3 dominate the time complexity. Step 3 is the Maximal Clique Enumeration Problem which is NP-Hard. One of the most common algorithms for this problem is the Improved BK by Bron and Kerbosch [3], whose worst-case time complexity has recently been proved to be $O(3^n/3)$ [14]. We use a variant of this algorithm with similar exponential complexity. To allow the algorithm to scale to a large number of robots, a quick calculation is done whenever the number of neighboring

[3] The idea behind the Clique Intensity Algorithm was proposed by Steven Damer.

Loop:
1. Update connectivity graph from neighbors' shared knowledge.
2. Share edges incident on me with neighbors.
3. Find all maximal cliques that I am in.
4. Determine the sentry for each maximal clique.
5. Choose and apply behavior based on sentries, cliques, and connectivity graph.

Behaviors:
1. Avoid Collisions Behavior:
   Utilize proximity sensors to avoid collisions
2. Seek Connection Behavior:
   Go in reverse for a bit and if this doesn't work pivot and move forward
3. Disperse Behavior:
   if my *sentry_intensity* is decreasing over time then go straight
   otherwise pivot for a bit and then move forward, and
   check for decreasing *sentry_intensity* again
4. Guard Behavior: Don't move.

Fig. 1. The high-level steps that roughly describes the Clique Intensity Algorithm. One of the four behaviors is applied each time through the loop.

robots exceeds a threshold. This quick calculation skips steps 1, 3, and 4, and instead each robot executes the disperse behavior and uses the neighbor with the lowest id as its sentry (Step 2 of Figure 2). See the discussion section for further explanation of this approach.

Step 1 of Figure 1, updating the connectivity graph, is the other dominant factor in the time complexity of this algorithm. Sharing the entire connectivity graph would provide the nice property that each robot has full knowledge of the connectivity graph across the swarm. Consider a scenario in which n robots begin close together such that the connectivity graph is fully connected, in which case the number of edges in the graph is $n*(n-1)/2$. Each robot would receive $n-1$ graphs from their neighbors, and must process each one to update its own knowledge. For this worst case scenario, the complexity in terms of the number of robots in the swarm becomes $O((n-1)*(n-1)*n/2) = O(n^3)$. In practice, when sharing the entire connectivity graph it becomes computationally difficult to have more than 20 robots in the swarm.

However, a robot does not need full knowledge of the connectivity graph for the Clique Intensity Algorithm to operate. There are two types of edges that are used by the algorithm. The first type are those that are incident on the robot, which are automatically given from the robot's wireless card. The second type are those that connect two robots which are both adjacent to the robot. These edges must be shared between robots. This is accomplished by each robot sharing only those edges incident on them to their adjacent neighbors. With a fully connected graph with n nodes, each robot receives

Loop:
1. If I am within proximity of a wall then apply Avoid Collisions Behavior
2. If number of neighbors is greater than *neighbor_threshold* then apply
   Disperse Behavior while using the neighbor with lowest id as my sentry.
3. Update connectivity graph from neighbors' shared knowledge
4. Share edges incident on me with neighbors.
5. If completely disconnected from all robots then apply Seek Neighbor Behavior
6. Find all maximal cliques that I am in.
   Two robots are connected if $signal\_intensity > \frac{1}{clique\_distance^2}$
7. For each maximal clique, choose a sentry
   A robot is a sentry at time i for clique j if:
      It is a sentry for another clique OR
      of all robots in clique j, it is in the most cliques OR
      if it is tied for being in the most cliques and has the lowest id in clique j
8. If I am a sentry then apply Guard Behavior
9. If I am an explorer, apply Disperse Behavior
   I am an explorer if:
      I am not a sentry AND
      my id is higher than at least 3 other robots in each of my cliques AND
      I am connected to more than 1 other robot.
10. Choose the sentry with the greatest intensity to be my sentry. Set *sentry_intensity* to be equal to the signal intensity between me and my sentry.
11. If $sentry\_intensity > \frac{1}{(clique\_distance-1)^2}$ then apply Disperse Behavior
12. Else apply Guard Behavior

Fig. 2. The detailed Clique Intensity Algorithm, from the perspective of a single robot in the swarm, hence the use of "I". Each time through the loop one of the four behaviors from Figure 1 is chosen and applied.

$n-1$ edges from $n-1$ adjacent neighbors, or a worst-case time complexity of $O(n^2)$. In practice this approach allows the algorithm to operate fluidly with 100 robots in the swarm.

A crucial aspect of the Clique Intensity Algorithm is how sentries are chosen. We would like as few robots as necessary to be sentries, so if a robot is a sentry for one clique, then it is considered a sentry for all of its cliques. If a robot on the perimeter becomes a sentry, then this will cause other robots to turn around and head back towards what is likely a more densely populated area. Therefore we want to encourage robots on the perimeter of the dispersion effort to continue dispersing. This is done by choosing the robot in a clique which is in the most cliques to be the sentry. This follows the idea that robots on the perimeter will be in fewer cliques. The tie-breaker for this situation is to simply choose the robot with the lowest id.

Creating cliques in the connectivity graph of size three or two is accomplished by designating certain robots as explorers. Given that they are still within range of at least one other robot, explorers will continue on out of range of their sentries. Any robot that is not an explorer or sentry will continue moving in direction of decreasing sentry intensity until a certain threshold is reached and will then stop.

The desired clique distance is an important parameter that can be supplied as input to the algorithm. The algorithm tries to create cliques in which the robots are separated by a distance approximated by the clique distance. Depending on the problem at hand, it may be desirable to set the clique distance as high as the expected wireless connectivity range. In other situations, such as a map with small rooms, we may want the clique distance to be considerably smaller than the wireless connectivity range.

4 Simulation Experiments

Simulations were ran using the Player/Stage robot server and two- dimensional simulation environment [5]. Player/Stage is probably the highest fidelity, most realistic, robot simulation software in the world. Robot clients written for the Player server originally for simulation are often easily transferred to a real robot with little or no modification necessary. Player/Stage does not facilitate simulation of wireless message passing. The additional simulation infrastructure built for this experiment includes simulating wireless connectivity and message passing amongst robots via TCP/IP sockets, an automatic launching system to facilitate running the simulation in a distributed manner, and an embedded web server in each robot to track the state of the simulation. The wireless signal intensity was simulated by a direct calculation of the distance between each robot, using the inverse square law of signal propagation ($Intensity = \frac{1}{(distance\ traveled\ by\ signal)^2}$). The effect of the environment, walls in particular, was not taken into consideration.

The metrics tracked during each experiment include initial area coverage, final area coverage, and total running time. Each robot is considered to cover an area within a specified radius of the robot. The area coverage is computed in a post-processing step using a script that takes as input the number of meters per pixel, radius, and an image of the final dispersion map. The number of pixels within the radius of each robot, taking walls into consideration, is counted and then multiplied by the resolution of a single pixel to produce as output the area coverage in square meters and an image depicting the coverage. The maximum velocity allowed was 0.3 meters per second.

4.1 Experiment A

A simple experiment with 12 robots was performed on a cave-like map. The clique distance was set at 5, and it was assumed the robots covered an area within a radius of 2.5 meters. The robots disperse from the starting configuration shown in Figure 3 until the algorithm reaches an equilibrium in which the entire swarm is stationary. The initial area coverage is 25 square meters.

This experiment was ran 10 times and resulted in a mean area coverage of 128 square meters with a standard deviation of 6.27 square meters. It took between 60 and 100 seconds for the algorithm to reach an equilibrium. Notice there are no disconnections in the area coverage.

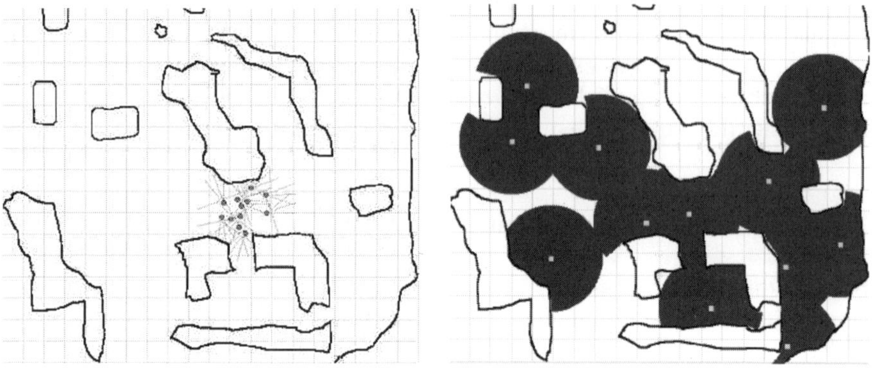

Fig. 3. Experiment A. Left: Starting configuration of experiment with 12 robots in a cave-like setting. A total of 25 square meters is covered at the start. Lines emanating from the robots indicate the range of the proximity sensors. Right: The output of the post-processing area coverage step for one run. Each robot covers an area within a 2.5 meter radius. This run resulted in an area coverage of 141 square meters in 80 seconds.

4.2 Experiment B

In this experiment, 100 robots started densely packed in the middle of a complicated floor of a hospital. The starting location is indicated by the light gray circle in Figure 4. The robots dispersed for 5 minutes and the clique distance was set at 5. It was assumed the robots covered an area within a radius of 4.5 meters. The algorithm was terminated after 5 minutes, even though an equilibrium was not achieved. The initial area coverage was 118 square meters. This experiment was ran 10 times and the mean area coverage was 1023 square meters with a standard deviation of 48 square meters. The is an improvement on average by a factor of 8.7.

5 Discussion

When a large number of robots are densely packed in a small area it is computationally too demanding for each robot to calculate cliques and share knowledge with its neighbors. The quick calculation done in step 2 of Figure 2 avoids this problem when the number of neighbors is above a threshold. Although

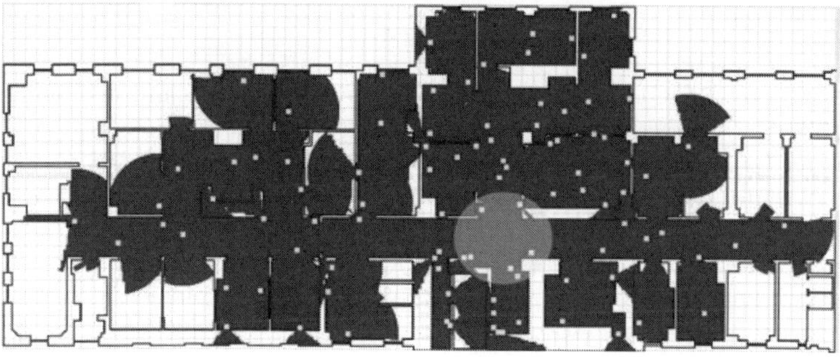

Fig. 4. Experiment B. 100 robots started in the light gray circle in the middle of the hospital map and dispersed outwards for 5 minutes. Each robot covers an area within a 4.5 meter radius. The total area covered for this particular run is 1063 square meters. The initial area coverage was 118 square meters.

this is necessary for a large number of robots, this causes inconsistencies that would not exist otherwise. As the swarm spreads out, some robots will switch modes to calculating cliques, while some will still be following the simpler approach. This means it is possible to have a situation in which one robot's sentry is moving (ex. 1 is connected to 2 and 2 is connected to 3). A way to avoid this problem would be to allow a robot to tell its sentry that it is using it as a sentry, in which case the sentry would know to stay put. Further experimentation is needed to verify this idea.

6 Conclusions and Future Work

The results obtained from both experiments indicate that a swarm of small robots can be dispersed effectively through the use of wireless signal intensities, without knowing the relative locations of neighboring robots. Using small robots instead of larger ones capable of carrying a laser range finder means that many more robots can be deployed for the same cost. It would be interesting to analyze how swarms with limited sensing abilities compare to swarms a fraction of their size with more powerful sensing abilities.

There is more work to be done on this topic. Variations in signal intensity should be modeled in a more realistic manner, by taking wall affects into consideration and applying gaussian noise. Eventually, experiments with real robots must be done to fully verify the potential of using wireless signal intensity in robot dispersion.

Acknowledgments

Partial funding provided by the National Science Foundation Industry/University Safety, Security, and Rescue Research Center, and by the Air Force under Contract No. FA8651-04-C-0191 to Architecture Technology Corporation.

References

1. M. Batalin and G. S. Sukhatme. Coverage, exploration and deployment by a mobile robot and communication network. *Telecommunication Systems Journal, Special Issue on Wireless Sensor Networks*, 26(2):181–196, 2004.
2. M. A. Batalin and G. S. Sukhatme. Spreading out: A local approach to multi-robot coverage. In *Proc. Int'l Symp. on Distributed Autonomous Robotic Systems"*, Fukuoka, Japan, 2002.
3. C. Bron and J. Kerbosch. Algorithm 457: Finding all cliques of an undirected graph. In *Proc. of the ACM*, 1973.
4. D. W. Gage. Command control for many-robot systems. In *19th Annual AUVS Technical Symposium*, pages 22–24, Huntsville, Alabama, June 1992.
5. B. P. Gerkey, R. T. Vaughan, K. Stöy, A. Howard, G. S. Sukhatme, and M. J. Matarić. Most valuable player: A robot device server for distributed control. In *Proc. IEEE/RSJ Int'l Conf. on Intelligent Robots and Systems*, pages 1226–1231, Oct. 2001.
6. B. Grabowski. Small robot survey. Carnegie Mellon University, http://www.andrew.cmu.edu/user/rjg/webrobots/small_robots_metric.html, 2003.
7. A. Haeberlen, E. Flannery, A. M. Ladd, A. Rudys, D. S. Wallach, and L. E. Kavraki. Practical robust localization over large-scale 802.11 wireless networks. In *Int'l Conf. on Mobile Computing and Networking (MOBICOM)*, pages 70–84, Philadelphia, PA, 2004.
8. T. He, C. Huang, B. M. Blumi, J. A. Stankovic, and T. Abdelzaher. Range-free localization schemes for large scale sensor networks. In *Int'l Conf. on Mobile Computing and Networking (MOBICOM)*. ACM, 2003.
9. A. Howard, M. J. Mataric, and G. S. Sukhatme. Mobile sensor network deployment using potential fields: A distributed, scalable solution to the area coverage problem. In *Proc. Int'l Symp. on Distributed Autonomous Robotic Systems"*, 2002.
10. T.-R. Hsiang, E. Arkin, M. A. Bender, S. Fekete, and J. Mitchell. Algorithms for rapidly dispersing robot swarms in unknown environments. In *Proc. 5th Workshop on Algorithmic Foundations of Robotics (WAFR)*, 2002.
11. B. Kadrovach and G. Lamont. Design and analysis of swarm-based sensor systems. In *Proc. of the Midwest Symposium on Circuits and Systems*, Aug. 2001.
12. P. E. Rybski, S. A. Stoeter, M. Gini, D. F. Hougen, and N. Papanikolopoulos. Performance of a distributed robotic system using shared communications channels. *IEEE Trans. on Robotics and Automation*, 22(5):713–727, Oct. 2002.
13. W. M. Spears, R. Heil, D. F. Spears, and D. Zarzhitsky. Physicomimetics for mobile robot formations. In *Autonomous Agents and Multi-Agent Systems*, 2004.
14. E. Tomita, A. Tanaka, and H. Takahashi. The worst-case time complexity for generating all maximal cliques. In *Proc. of Computing and Combinatorics Conference*, 2004.

Distributed, Play-Based Role Assignment for Robot Teams in Dynamic Environments

Colin McMillen and Manuela Veloso

School of Computer Science, Carnegie Mellon University, Pittsburgh, PA, U.S.A.
{mcmillen,veloso}@cs.cmu.edu

Summary. The design of a coordination strategy for a distributed robotic team is challenging in domains with high uncertainty and dynamic environments. We present a distributed, play-based role assignment algorithm that has been implemented on real robots in the RoboCup four-legged league. The algorithm allows the robots to adapt their strategy based on the current state of the environment, the game, and the behavior of opponents. The distributed play-based approach also enables the robots to reason about task-based temporal constraints and has been designed to be resistant to the problem of role oscillation.

1 Introduction

A common goal of distributed autonomous robotic systems is the development of teamwork and coordination strategies. The benefits of adding multiple robots to a system, such as increased performance and reliability, have been demonstrated in many different situations. However, depending on the domain and the task, different sorts of approaches might be needed. We are interested in the implementation of multi-robot coordination in domains with high uncertainty and dynamic environments.

In this paper, we present our approach to *role assignment* in the RoboCup four-legged league [2], in which two teams of four Sony AIBO robots compete in a robot soccer game. Figure 1 shows a snapshot of a recent AIBO game. This domain presents many challenges, including: full robot autonomy, distributed robot team control, limited individual robot perception, the presence of robot adversaries, task-dependent temporal constraints, and high communication latency. In this paper, we contribute a new distributed *play-based* system that equips the robots with *plays* – alternative teamwork strategies. This method was developed to overcome limitations of previous approaches. In particular, it assigns roles to robots in a fault-tolerant manner that minimizes role switching and synchronization problems. Our approach has been fully implemented within robot soccer, but is designed to be relevant to general multi-robot domains that share some of the challenging features of robot soccer, as identified in this paper.

Fig. 1. A RoboCup four-legged league soccer match.

We discuss the RoboCup legged league in section 2, identifying the technical features that we address as especially challenging for multi-robot coordination. Section 3 introduces our approach to teamwork. Section 4 presents experimental results that highlight some of the advantages of our approach. We present our conclusions in section 5. Related work is discussed throughout the paper as needed.

2 Challenges of the RoboCup Legged League

In the RoboCup four-legged league, two teams of four Sony AIBO robots play each other in a time-limited and space-limited setting (currently two game halves of ten minutes each and a field of 6m×4m). The settings and the rules of the game change every year to create new research challenges. The current complete rules of the domain are available at [2]. We focus on the discussion of the general features of the game that are of relevance to team coordination. These features include:

1. **Full autonomy:** each team of robots operates completely without human supervision. However, teams are allowed to change the robots' programming at halftime or during a timeout. Each team is granted one timeout per game.
2. **Distributed teams:** all perception, computation, and action is done onboard the robots. The robots are equipped with 802.11b wireless networking, which enables communication among team members; however, the robots are not allowed to communicate with any off-board computers.

3. **Limited perception:** each robot's primary sensor is a low-resolution camera with a very narrow field of view (under 60 degrees). A single robot therefore has a very limited view of the world, so teams can benefit greatly from communication strategies that build a shared world model.
4. **Dynamic, adversarial environment:** the presence of adversaries in the environment is a significant challenge. Opponents ensure that the environment is extremely dynamic: within a few seconds, the state of the world may change significantly.
5. **Temporal constraints:** there are two temporal constraints that arise due to the presence of adversaries. First, all team decisions must occur in real time. A team that takes too long to coordinate will have robots that display hesitation in carrying out their tasks, which gives the opponents a significant advantage. Second, soccer is a finite-horizon zero-sum game. A game of soccer has a winning team, a losing team, and a defined ending point. Playing a conservative strategy – which might work well over a long period of time – is of no use to a team that is losing and only has a few seconds remaining in the game. A team in this situation must choose a strategy that can score a goal quickly, even if such a strategy has other weaknesses. Multiple team coordination strategies are needed. The selection algorithm faces complex multi-objective optimization criteria.
6. **High network latency:** the presence of dozens of robots in the competition environment leads to very unpredictable quality of the robots' wireless network. Teams may experience periods of high network latency and collisions; latencies of over a second have been observed. To achieve consistent performance, a team needs to ensure that the coordination strategies employed are robust to disruptions in communication.

The issues of limited perception in the four-legged league have already been addressed in previous work. Our specific implementation of a shared world model [7] is not discussed at length here, but it is important to note that some of the decisions made by individual robots on our team are influenced by information that has been communicated by teammates.

In this paper, we focus on the challenges of role assignment (also known as *task allocation*) in the RoboCup legged league. Due to the constraints of our domain, it is a requirement that the solution be implemented on fully autonomous, distributed robots. The work presented in this paper specifically addresses the issues of adversarial environments, temporal constraints, and robustness under high network latency.

Gerkey and Mataric [4] discuss role assignment in RoboCup, giving an overview of the strategies used by teams in each of the RoboCup leagues. In the four-legged league, the predominant approach involves allocating three roles: an *attacker*, a *defender*, and some sort of *supporter*. According to Gerkey's survey, nearly all RoboCup teams assign roles to robots in a greedy fashion. For example, in previous years, our own team (CMPack) assigned the roles in a fixed order using a well-defined objective function, namely first the *attacker* role to the robot that could reach the ball most quickly, then the *defender* role to whichever of the other two robots was closest to our own goal, and finally the *supporter* role to the remaining robot [8]. To prevent robots from interfering with one another, the attacker was the only robot allowed to actually approach the ball for a kick; the other robots would position themselves in useful supporting locations. If the ball came near to another robot, the team members would negotiate a role switch. After the role switch, the closest robot to the ball would become the new attacker, and the attack would continue. This was a very effective strategy that contributed strongly to our world championship in 2002. However, this strategy has some limitations, particularly in terms of role switching and oscillation. Two robots that are equally suited to the attacker role might fight over it. The potential outcome is an undesirable role oscillation between the robots – a period of hesitation where neither robot makes significant progress toward completing the task. A standard solution to this problem, which has previously been implemented in our own team and other RoboCup teams, is to add some hysteresis to the role assignment: either allowing role assignments to occur infrequently (e.g., every few seconds) or ensuring that a role switch will not occur unless one robot is significantly more suited to the task than another. However, adding hysteresis to a system can be difficult: if too much hysteresis is added, the robots will miss out on opportunities because they are no longer reacting as quickly to changes in their dynamic environment. In practice, finding a solution that balances these two constraints (lack of role oscillation and immediate response to environmental changes) can be difficult. This problem is exacerbated under the presence of communication failures or high network latency, where the negotiation of a role switch may take several seconds to complete. Another limitation of this design is that there can be a significant cost to role switching, as the robots reconfigure themselves for their new roles. The work presented in this paper attempts to minimize the effect of these limitations.

Dylla *et al.* [3] propose a soccer strategy language that formalizes the strategies and tactics used by human soccer teams. Their goal is to be able to specify soccer strategies in an abstract way that does not depend strongly on the specific robot hardware used in the competition. However, to our knowledge, the authors do not yet plan to use this language in the implementation of any real robot soccer team. In this paper, we present a strategy language that could be used in any multi-robot system and that has been implemented in the four-legged league of the 2005 RoboCup competition.

3 Distributed Play-Based Role Assignment

We can say that teamwork in general consists of a team control policy, i.e., a selection of a joint action by teammates given a perceived state of the environment [6]. It is our experience that it is rather challenging to generate or learn a team control policy in complex, highly dynamic (in particular adversarial), multi-robot domains. Therefore, instead of approaching teamwork in terms of a mapping between state and joint actions, we follow a *play-based* approach, as introduced by Bowling *et al.* [1]. We introduce a distributed play-based approach to teamwork, which allows us to handle the domain challenges introduced in section 2. A play specifies a *plan* for the team; i.e., under some applicability conditions, a play provides a sequence of steps for the team to execute. Multiple plays can capture different teamwork strategies, as explicit responses to different types of opponents. Bowling showed that play selection weights could be adapted to match an opponent. Plays also allow the team to reason about the zero-sum, finite-horizon aspects of a game-playing domain: the team can change plays as a function of the score and time left in the game. Our play-based teamwork approach ensures that robots do not suffer from hesitation nor oscillation, and that team performance is not significantly degraded by possible periods of high network latency. We believe that ours is the first implemented distributed play-based teamwork approach, at least in the context of the RoboCup four-legged league.

3.1 Plays

A *play* is a team plan that provides a set of roles, which are assigned to the robots upon initiation of the play. Bowling [1] introduced a play-based method for team coordination in the RoboCup small-size league. However, the small-size league has centralized control of the robots. One of the significant contributions of our work is the development of a play system that works in a distributed team. The play language described by Bowling assumes that the number of robots is fixed, and therefore always provides exactly four different roles for the robots. In another extension to Bowling's work, our plays also specify which roles are to be used if the team loses some number of robots due to penalties or crashes. This extension to the role-assignment aspects of Bowling's play language allows the team to robustly adapt to the loss of team members without the need for additional communication. This is a particularly important extension for domains where limited or high-latency communication is the norm.

Our play language itself is also strongly inspired by the work of Bowling. Our language allows us to define *applicability conditions*, which denote when a play is suitable for execution; what *roles* should be assigned when we have a specific number of active robots on the team; and a *weight*, which is used to decide which play to run when multiple plays are applicable.

Applicability. An applicability condition denotes when a play is suitable for execution. Each applicability condition is a conjunction of binary predicates. A play may specify multiple applicability conditions; in this case, the play is considered executable if any of the separate applicability conditions are satisfied.

Roles. Each play specifies which roles should be assigned to a team with a variable number of robots by defining different `ROLES` directives. A directive applies when a team has k active robots, and specifies the corresponding k roles to be assigned. If a robot team has n members, each play has a maximum of n `ROLES` directives. Since our AIBO teams are composed of four robots, our plays have four `ROLES` directives.

Weight. Weight is used to decide which play to run when multiple plays are applicable. In our current implementation, the play selector always chooses the applicable play with greatest weight. Future work could include choosing plays probabilistically based on the weight values or updating the weights at execution time to automatically improve team performance. *Playbook adaptation* of this sort has been implemented by Bowling for the small-size league [1]. In the work presented in this paper, adaptation was not used – the weights were set manually.

Unlike the work of Bowling, we do not have `DONE` or `TIMEOUT` keywords that specify when a play is complete. Rather, a play is considered to be complete as soon as the play selector chooses a different play, which may happen because the current play is no longer applicable or because another play with greater weight has recently become applicable. Each predicate used in an applicability condition is designed with some hysteresis, such that it is not possible for the predicate to rapidly oscillate between true and false. The predicates used in our approach depend on features of the environment – such as the time left in game, the number of goals scored by each team, and the number of robots available to each team – that by their nature cannot rapidly oscillate. This ensures that the play choice also cannot rapidly oscillate.

Figure 2 shows an example of a defensive play. Its applicability conditions specify that this play is applicable 1) when our team has fewer active players than the opponents or 2) when the game is in the second half and our team is winning by at least two points. If we have only one active robot on our team, we will assign it the Goalkeeper role; if we have two robots, one is assigned the Goalkeeper role and the other is assigned the Defender role; and so on.

```
PLAY StrongDefense
APPLICABLE fewerPlayers
APPLICABLE secondHalf winningBy2OrMoreGoals
ROLES 1 Goalkeeper
ROLES 2 Goalkeeper Defender
ROLES 3 Goalkeeper Defender Independent
ROLES 4 Goalkeeper Defender Midfielder Independent
WEIGHT 3
```

Fig. 2. An example play with multiple applicability conditions.

3.2 Play Selector

The *play selector* runs continuously, on one robot that is arbitrarily chosen to be the leader. The play selector chooses which play the team should be running. The leader periodically broadcasts the current play (and role assignments) to its teammates. Distributed play-based coordination is achieved through a predefined agreement among the team members to resort to a *default play* if a robot doesn't hear a play broadcast within a *communication time limit.* A failure of the leader or a network problem may trigger this default coordination plan. The algorithm used by the play selector is presented in Figure 3.

```
SELECT_PLAY(S: world state, P: playbook, D: default play):
  BEST_PLAY <- D
  BEST_WEIGHT <- WEIGHT(D)
  for each PLAY in P:
    if WEIGHT(PLAY) > BEST_WEIGHT:
      for each CONDITIONS in APPLICABLE(PLAY):
        if all CONDITIONS are satisfied in STATE:
          BEST_PLAY <- PLAY
          BEST_WEIGHT <- WEIGHT(PLAY)
  return BEST_PLAY
```

Fig. 3. Algorithm used by the play selector.

3.3 Roles

The *role* assigned to each robot determines what behaviors the robot actually runs. Our approach, used in RoboCup 2005, is unique in that it is *region-based*: each robot is assigned to a region of the field. A robot is primarily responsible for going after the ball whenever the ball is in that robot's region. When the ball is not in its region, the robot will position itself at a good position within its region (defined by a role-dependent objective function). Unlike previous approaches, robots no longer need to negotiate with one another in order to gain the *attacker* role that allows them to approach the ball. In this way, the performance of the team does not degrade significantly under high network latency. To ensure that one or more robots are always chasing after the ball, these regions typically overlap significantly. We have developed algorithms that prevent the robots from interfering with one another even when they are playing in overlapping regions. To provide robustness against communication failure, these algorithms are designed to operate without the need for communication, using local information such as a robot's vision of its own teammates. However, if communication is available, we can use additional features (such as reported teammate positions and ball positions) that provide additional confidence that our robots will not interfere with one another.

3.4 Role Allocation

The selection of a play determines which roles need to be allocated to the robots. However, it does not specify which robots should be assigned to each role. Therefore, a role allocation algorithm is still needed to assign the roles. This algorithm also runs on the leader robot, which broadcasts the assignment along with the selected play. Our role allocator has two features that differentiate it from those used by most other RoboCup legged-league teams [4]. First, it only runs when a play is initially selected, as opposed to continuously. Second, it allocates roles in a *role-preserving* manner – minimizing role switching. Formally, if a new play P_t is selected at time t, and P_t specifies n roles $\{R_{1..n}\}$ for the n robots $r_{1..n}$, and r_i was already assigned to R_j in P_{t-1}, r_i is guaranteed to still be assigned to R_j in P_t. (Any remaining roles can be allocated in a greedy fashion.) The region-based nature of our plays enables this feature, as r_i is already guaranteed to be in the region for R_j since it was already in that role's region in the previous play. Therefore, it can assume the new role without any transitional cost. These features provide additional resistance to oscillation in cases in which two plays share common roles.

4 Experimental Results

We have previously presented empirical results that support the feasibility and effectiveness of multiple plays in the RoboCup four-legged league [5]. In this paper, we contribute a role allocation strategy, claiming that it addresses hesitation due to role oscillation by preserving a robot's role when possible. We show experimental evidence that supports this particular claim.

In each experimental trial, three robots work together in a robot soccer task, namely *ball advancement* – moving the ball towards the opposing goal as quickly as possible. Figure 4 shows the initial position of the robots, from which the team advances the ball down the field towards the goal. A trial is considered complete when either a goal is scored, the ball advances past the opponents' back line, or the ball hits one of the goal posts. The time of each completed trial is measured.

We test the robots' teamwork in three different team play configurations: (i) a single *Defender-Striker-Independent* play; (ii) a single *Defender-Midfielder-Independent* play; and (iii) switching every five seconds between the two plays in (i) and (ii). Since these two plays share two roles (defender and independent), we expect that, even with frequent play switching, our role assignment algorithm will not adversely affect the performance of the team.

Each configuration was tested for 40 completed trials, for a total of 120 experiments. Figure 5 summarizes the results. The fastest and slowest times achieved in any trial were 17.18 and 70.15 seconds, respectively. The *Defender-Striker-Independent* play performs best at this task, completing each trial in a mean time of 31.06 seconds. The *Defender-Midfielder-Independent* play

Fig. 4. Initial position for each experimental trial. The three robots are placed in three positions on the field, with the ball in the defense area. The experiment proceeds until the robots advance the ball past the end line of the opposite half of the field.

performs more slowly, completing each trial in a mean time of 35.05 seconds. The difference between these times is significant (determined by Student's two-tailed t-test, with $p = 0.048$). When the robots oscillate between these two plays, their performance remains good, with the mean time of the switching case (33.29 seconds) between the mean times of the other two cases. Since the play-switching case still performs better than the worse of the two plays, we note that there is no significant detrimental effect on performance.

5 Conclusion

In this paper, we have presented the details of a distributed play-based role assignment algorithm, which has been implemented on a distributed team of robots for the RoboCup four-legged league. The algorithm aims to solve several important general distributed multi-robot challenges, including the presence of adversaries, task-based temporal constraints, and robustness to network failure. We have presented experimental results that show that our role-preserving assignment algorithm allows a team to perform well even when plays are rapidly changed.

The presented role-assignment algorithm and plays have been tested in the RoboCup 2005 competition. Our team came in fourth place in a challenging competition of twenty-four teams. Our team typically rotated through three well-balanced plays in the first minutes of each game, which allowed us to see the performance of each play against the specific opponent. As a form of adjustable autonomy, we could manually change the team's strategy at halftime or during a timeout. Our future work includes the investigation

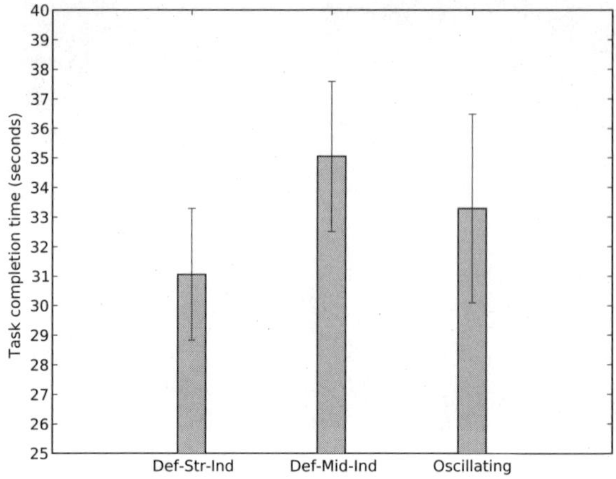

Fig. 5. Experimental results for the *Defender-Striker-Independent* play, *Defender-Midfielder-Independent* play, and switching between the two plays. The figure shows the means and 90% confidence intervals for each case.

of automatic play adaptation in distributed domains and recognition of the opponents' state and strategy.

References

1. M. Bowling, B. Browning, A. Chang, and M. Veloso. Plays as team plans for coordination and adaptation. In D. Polani, B. Browning, A. Bonarini, and K. Yoshida, editors, *RoboCup 2003: Robot Soccer World Cup VII*. 2004.
2. RoboCup Tech. Committee. Four legged robot football league rule book, 2006.
3. F. Dylla, A. Ferrein, G. Lakemeyer, J. Murray, O. Obst, T. Röfer, F. Stolzenburg, U. Visser, and T. Wagner. Towards a league-independent qualitative soccer theory for RoboCup. In *RoboCup 2004: Robot Soccer World Cup VIII*. 2005.
4. B. Gerkey and M. Mataric. On role allocation in RoboCup. In D. Polani, B. Browning, A. Bonarini, and K. Yoshida, editors, *RoboCup 2003: Robot Soccer World Cup VII*. 2004.
5. C. McMillen, P. Rybski, and M. Veloso. Levels of multi-robot coordination for dynamic environments. In *Multi-Robot Systems: From Swarms to Intelligent Automata, Volume III*. 2005.
6. D. Pynadath and M. Tambe. The communicative multiagent team decision problem: Analyzing teamwork theories and models. *Journal of AI Research*, 2002.
7. M. Roth, D. Vail, and M. Veloso. A real-time world model for multi-robot teams with high-latency communication. In *Proc. IEEE/RSJ Intl. Conf. on Intelligent Robots and Systems (IROS)*, 2003.
8. D. Vail and M. Veloso. Dynamic multi-robot coordination. In *Multi-Robot Systems: From Swarms to Intelligent Automata, Volume II*. 2003.

Simultaneous Planning, Localization, and Mapping in a Camera Sensor Network

David Meger[1], Ioannis Rekleitis[2], and Gregory Dudek[1]

[1] McGill University, Montreal, Canada [dmeger,dudek]@cim.mcgill.ca

[2] Canadian Space Agency, Longueuil, Canada Ioannis.Rekleitis@space.gc.ca

Summary. In this paper we examine issues of localization, exploration, and planning in the context of a hybrid robot/camera-network system. We exploit the ubiquity of camera networks to use them as a source of localization data. Since the Cartesian position of the cameras in most networks is not known accurately, we consider the issue of how to localize such cameras. To solve this hybrid localization problem, we subdivide it into a local problem of camera-parameter estimation combined with a global planning and navigation problem. We solve the local camera-calibration problem by using fiducial markers embedded in the robot and by selecting robot trajectories in front of each camera that provide good calibration and field-of-view accuracy. We propagate information among the cameras and the successive positions of the robot using an Extended Kalman filter. The paper includes experimental data from an indoor office environment as well as tests on simulated data sets.

1 Introduction

In this paper we consider interactions between a mobile robot and an emplaced camera network. In particular, we would like to use the camera network to observe and localize the robot, while simultaneously using the robot to estimate the positions of the cameras (see Fig. 1a). Notably, networks of surveillance cameras have become very commonplace in most urban environments. Unfortunately, the actual positions of the cameras are often known only in the most qualitative manner. Furthermore, geometrically accurate initial placement of cameras appears to be inconvenient and costly. To solve this hybrid localization problem, we will divide it into two interconnected sub-problems. The first is a local problem of camera-parameter estimation which we solve by using fiducial markers embedded in the robot and by selecting robot trajectories before each camera that provide good calibration and field-of-view accuracy. The second problem is to move the robot over large regions of space (between cameras) to visit the locations of many cameras (without *a priori* knowledge of how those locations are connected). That, in turn, entails uncertainty propagation and planning.

In order for the camera network and the robot to effectively collaborate, we must confront several core sub-problems:

1. **Estimation -** detecting the robot within the image, determining the camera parameters, and producing a metric measurement of the robot position in the local reference frame of the camera.

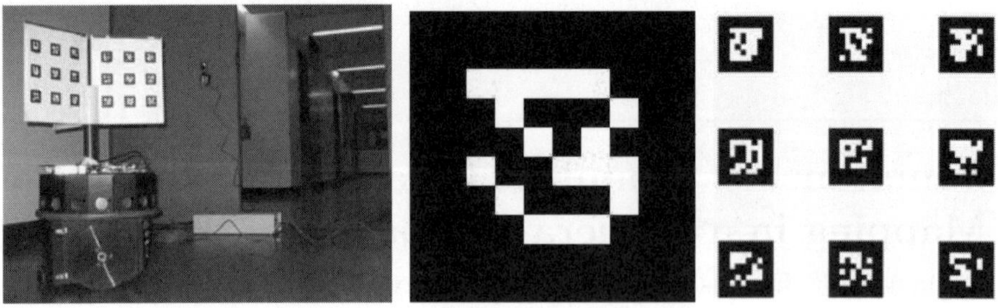

(a) (b) (c)

Fig. 1. (a) The robot with calibration patterns on the target in front of a camera. (b) An example ARTag marker. (c) A calibration target formed from ARTag markers

2. **Local planned behavior -** planning the behavior of the robot within the field of view of a single camera, the robot needs to facilitate its observability and, ideally, maximize the accuracy of the camera calibration.
3. **Data fusion -** combining local measurements from different sources in order to place the cameras and the robot in a common global frame.

The task of computing camera parameters and obtaining metric measurements is referred to as *camera calibration* and is well-studied in both photogrammetry and computer vision [6, 1]. Typically camera calibration is a human intensive task. Section 3.1 will detail an automated version where the robot replaces the human operator in moving the calibration pattern. A system of bar-code-like markers (see Fig. 1) is used along with a detection library [7] so that the calibration points are detected robustly, with high accuracy, and without operator interaction.

Measurements from the calibration process can be used to localize the robot and place each camera within a common reference frame. This process can be formulated as an instance of Simultaneous Localization and Mapping (SLAM). Typically the robot uses its sensors to measure the relative locations of landmarks in the world as it moves. Since the measurements of the robot motion as well as those of the pose of landmarks are imperfect, estimating the true locations becomes a filtering problem, which is often solved by using an Extended Kalman filter (EKF). Our situation differs from standard SLAM in that our sensors are not pre-calibrated to provide metric information. That is, camera calibration must be performed as a sub-step of mapping.

The path that the robot follows in front of a single camera during calibration will allow a variety of images of the target to be taken. During this local exploration problem, the set of captured images must provide enough information to recover camera parameters. The calibration literature [22] details several cases where a set of images of a planar target does not provide sufficient information to perform the calibration. The robot must clearly avoid any such situation, but we can hope for more than just this simple guarantee. Through analysis of the calibration equations, and the use of the robot odometry, the system discussed here has the potential to perform the calibration optimally and verify the results.

The following section discusses related research. Section 3 details camera calibration using marker detection and a 6 degree of freedom (DOF) EKF for mapping in our context. Section 4 continues the discussion of local calibration paths. Section 5 provides experimental results to examine the effect of different local paths and shows the system operating in an office environment of 50 m in diameter. We finish this paper with concluding remarks.

2 Related Work

Previous work on the use of camera networks for the detection of moving objects has often focused on person tracking in which case the detection and tracking problem is much more difficult than that of our scenario (due to lack of cooperative targets and a controllable robot) [9, 4, 5]. Inference of camera network topology from moving targets has been considered [4, 13]. Ellis *et al.* depend on cameras with overlapping fields of view. Marinakis *et al.* deal with non-overlapping cameras, but only topological information is inferred here while we are interested in producing a metric map of the cameras. Batalin and Sukhatme [2] used the radio signals from nodes in a sensor network only for the localization of the robot. Cooperative localization among robots has been considered [11, 15, 17, 10], where instead of camera nodes a robot is observed by other robots.

Camera calibration is a well studied problem; a good summary paper by Tsai [19] outlines much of the previous work, and [22] presents improvements made more recently. A series of papers by Tsai et al. [21, 20] use a 3-D target and a camera mounted on the end of a manipulator to calibrate the manipulator as well as the camera. Heuristics are provided in [21] to guide the selection of calibration images that minimizes that error. However, these methods only deal with a single camera and use manipulators with accurate joint encoders, *i.e.*, odometry error is not a factor.

One important step in the automation of camera calibration is the accurate detection of the calibration pattern in a larger scene. Fiducial markers are engineered targets that can be detected easily by a computer vision algorithm. ARToolkit [14] and ARTag [7] are two examples. ARTag markers are square black and white patches with a relatively thick solid outer boundary and an internal 6 by 6 grid (see Fig. 1b,c). The advantages of this system are reliable marker detection with low rates of false positive detection and marker confusion. ARTag markers have been previously used for robot localization in [8] where a camera from above viewed robots, each of which had a marker attached on top.

The EKF is used for mapping in the presence of odometry error, a method that was detailed by [18],[12] and others to form the now very mature SLAM field. An example of previous use of camera networks for localization and mapping is [16]. Our work extends this previous method by using ARTag markers for much more automated detection of calibration target points, performing SLAM with 3-D position and orientation for cameras and examining the effect of local planning. This gives our system a higher level of autonomy and allows mapping of much larger environments.

3 Mapping and Calibration Methods

Our approach to the general problem of mapping a camera sensor network is divided into two sub-problems: acting locally to enhance the intrinsic parameter estimation; and moving globally to ensure coverage of the network while maintaining good accuracy. As it visits each location for the first time, the robot is detected by a camera. Thus, it can exploit its model of its own pose, and the relative position of the camera to the robot to estimate the camera position. In order to recover the coordinate system transformation between the robot and the camera, it is necessary to recover the intrinsic parameters of the camera through a calibration procedure. This process can be facilitated by appropriate local actions of the robot. Finally, over the camera network as a whole, the robot pose and the camera pose estimates are propagated and maintained using a Kalman filter.

A target constructed from 6 grids of AR-Tag markers is used for automated detection and calibration. When the robot moves in front of a camera, the markers are detected, and the corner positions of the markers are determined. A set of images is collected for each camera, and the corner information is used to calibrate the camera. Once a camera is calibrated, each subsequent detection of the robot results in a relative camera pose measurement. The following sub sections provide details about the steps of this process.

3.1 Automated Camera Calibration

A fully automated system is presented for the three tasks involved in camera calibration: collecting a set of images of a calibration target; detecting points in the images which correspond to known 3-D locations in the target reference frame; and performing calibration, which solves for the camera parameters through non-linear optimization. The key to this process is the calibration target mounted atop a mobile robot as shown in Fig. 1a. The marker locations can be detected and the robot can then move slightly, so that different views of the calibration targets are obtained until a sufficient number is available for calibration. Six panels, each with 9 markers, are mounted on three vertical metal planes. The 3-D locations of each marker corner in the robot frame can be determined through simple measurements.

The ARTag detection algorithm relies on identification of the fine internal details of the markers. This requires the marker to occupy a large portion of the image and limits the maximum detection distance to about 2 m in our setup. Of course higher-resolution camera hardware and larger calibration patterns will increase this distance.

The non-linear optimization procedure used for camera calibration [22] warrants a brief discussion. A camera is a projective device, mapping information about the 3-D world onto a 2-D image plane. A point in the world $M = [X, Y, Z]^T$ is mapped to pixel $m = [u, v, 1]^T$ in the image, under the following equation:

$$s\underbrace{\begin{bmatrix} u \\ v \\ 1 \end{bmatrix}}_{\mathbf{m}} = \underbrace{\begin{bmatrix} f_x & \alpha & u_x \\ 0 & f_y & u_y \\ 0 & 0 & 1 \end{bmatrix}}_{\mathbf{A}} \underbrace{\begin{bmatrix} R\, t \end{bmatrix}}_{\mathbf{T}} \underbrace{\begin{bmatrix} X \\ Y \\ Z \\ 1 \end{bmatrix}}_{\mathbf{M}} \tag{1}$$

In matrix A, f_x and f_y represent the focal lengths in pixel related coordinates, α is a skew parameter and u_x and u_y are the coordinates of the center of the image. Collectively, these are referred to as intrinsic camera parameters. s is a projective scale parameter. The T matrix is a homogeneous transformation made up of rotation R and translation t, and it expresses the position and the orientation of the camera with respect to the calibration-target coordinate frame. The elements of T are referred to as extrinsic parameters and change every time the camera or the calibration target moves to describe the position of the target relative to the camera. We will use the T matrix as a measurement in the global mapping process described in detail in Section 3.2.

The calibration images give a number of correspondences $(u, v) \rightarrow (X, Y, Z)$, which are related by (1). This relation allows the intrinsic camera parameters and the extrinsic parameters of each image to be jointly estimated using a two-step process. The first step is a linear solution to find the most likely intrinsic parameters. The second step is a non-linear optimization which includes polynomial distortion parameters. Zhang [22] mentions "degenerate configurations" where a set of calibration points do not provide enough information to solve for A and T. This occurs when all of the points lie in a lower dimensional linear subspace of R^3. To avoid this situation, several different local motion strategies are discussed in Section 4.

In conclusion, from a set of images of the robot-mounted target, the camera intrinsic and extrinsic parameters are estimated. The next section will discuss the use of an Extended Kalman filter to combine these estimates with robot odometry in order to build a map of camera positions.

3.2 Six-DOF EKF

The measurements of the extrinsic camera parameters can be used to build a consistent global map by adding the camera position to the map when initial calibration finishes and by improving the estimate each time the robot returns to the camera. To maintain consistent estimates in this global mapping problem, an Extended Kalman filter is used to combine noisy camera measurements and odometry in a principled fashion. The robot pose is modeled as position and orientation on the plane: (x, y, θ). However, the cameras may be positioned arbitrarily; so, their 3-D position and orientation must be estimated. Roll, pitch, and yaw angles are used to describe orientation, thus the state of each camera pose is a vector $X_c = [x, y, z, \alpha, \beta, \gamma]^T$ (for more information, see [3]).

The EKF tracks the states of the robot and the cameras in two steps: the propagation step tracks the robot pose during motion, and the update step

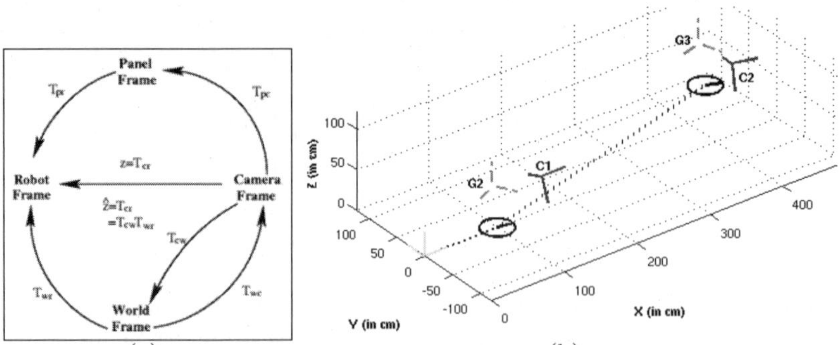

(a) (b)

Fig. 2. (a) Measurement Coordinate Frame Transformations. (b) The world (at [0,0,0]), robot (denoted by a circle and a line for the orientation), target grid (dashed lines G1,G2) and camera (solid lines C1,C2) coordinate frames. The trajectory of the robot is marked by a dotted line.

corrects the robot and the camera poses based on the measurements from the calibration process. For the propagation phase, the state vector and the covariance matrix are updated as in [18].

$$\hat{X}_{k|k-1} = F\hat{X}_{k-1|k-1} \tag{2}$$

$$P_{k|k-1} = FP_{k-1|k-1}F^T + C_v \tag{3}$$

where F is obtained by linearizing the non-linear propagation function $f(X,u)$ at state X and control actions u, and C_v is a matrix representing odometry error. For the update phase, the measurement equation is a non-linear expression of the state variables so we must again linearize before using the Kalman filter update equations. The measurement equation relates two coordinate frames, so that the language of homogeneous coordinates transformations is used in order to express the relation [3].

The calibration process estimates the calibration panel in the camera frame, that is ${}^C_P T$. Using ${}^P_R T$ which is measured initially this can be transformed into a relation between the camera and robot: ${}^C_R T = {}^C_P T {}^P_R T$. This is the measurement z. Next, the measurement is expressed in terms of the EKF states X_r and $X_c$through which we obtain the transformations for the robot and the camera in world coordinates: ${}^W_R T$ and ${}^W_C T$. Fig. 2 shows the relationships between the EKF state variables and the information obtained from camera calibration which jointly form the measurement equation:

$$z_{measured} = {}^C_R T = {}^C_W T {}^W_R T = {}^W_C T^{-1} {}^W_R T = \begin{bmatrix} {}^W_C R^T & -{}^W_C R^T {}^W_C P \\ 0 & 1 \end{bmatrix} \begin{bmatrix} {}^W_R R & {}^W_R P \\ 0 & 1 \end{bmatrix}$$

$$= \begin{bmatrix} {}^W_C R^T {}^W_R R & {}^W_C R^T ({}^W_R P - {}^W_C P) \\ 0 & 1 \end{bmatrix} \tag{4}$$

Equation 4 provides the measurement equation $\hat{z} = h(\hat{X})$. To use this in a Kalman filter, we must differentiate h with respect to each parameter to obtain a first-order linear approximation $z = h(\hat{X}) + H\tilde{X}$ where H is the Jacobian of vector function h. Measurement noise C_ω expresses the uncertainty of transformation parameters from camera calibration. The EKF update equations can be applied as usual:

$$\hat{X}_{k|k} = \hat{X}_{k|k-1} + K(z - h(\hat{X}_{k|k-1})) \tag{5}$$

$$P_{k|k} = \left[I - KH^T\right] P_{k|k-1} \tag{6}$$

$$K = P_{k|k-1} H (H P_{k|k-1} H^T + C_\omega)^{-1} \tag{7}$$

4 Local Calibration Procedures

Using a robot-mounted target provides a unique opportunity to collect calibration images in an intelligent fashion by controlling the robot motion. However, it is not immediately clear what the best motion strategy will be. There are numerous sources of error including detecting the original pixels, approximating the linear parameters, and convergence of the non-linear optimization all of which should be minimized if possible. As mentioned previously, [22] showed that it is essential to avoid having only parallel planes. [21] discussed heuristics for obtaining images to calibrate a manipulator system. Also, the accumulated odometric error is an important factor for the overall accuracy of the system.

As an initial investigation into this problem, five motion strategies were examined. These were chosen to cover the full spectrum of expected calibration accuracy and odometry buildup:

- **Stationary** - the robot moves in the camera field of view (FOV) and stays in one spot. Due to the target geometry, this allows for two non-parallel panels to be observed by the camera, which provides the minimal amount of information necessary for calibration.
- **One Panel Translation-only** - the robot translates across the camera FOV with only a single calibration panel visible always at the same angle. This is a degenerate case and produces inconsistent results.
- **Multi-Panel Translation-only** - the robot translates across the camera FOV with two panels visible. This provides numerous non-parallel planes for calibration and accumulates minimal odometry error.
- **Rotation-only** - the robot rotates in place in the center of the camera FOV allowing the panels to be detected at different angles.
- **Square Pattern** - the robot follows a square-shaped path in front of the camera. Since there is variation in the detected panel orientation and in depth, this method achieves good calibration accuracy. However, the combination of rotation and translation accumulates large odometry error.

5 Experimental Results

Two separate sets of experiments were conducted using the camera sensor network (see [16] for a detailed description of the experimental setup) which dealt with the mapping and the calibration aspects of our system. First, the five different local motion strategies were examined with respect to the resulting intrinsic parameters and position accuracy. Second, to show that mapping is feasible in a real-world environment, a robot equipped with the calibration target moved through one floor of an office building which was over 50 m in diameter. We show that the robot path estimate is improved through the use of position measurements from a set of cameras present in the environment.

5.1 Local Calibration Paths

A set of experiments was performed to test the effects of the local calibration paths suggested in Section 4. The goal was to study the motion strategies in terms of reliable camera calibration as well as magnitude of odometry error. This test was done inside our laboratory with a Nomadics Scout robot mounted with a target with six calibration patterns. The 5 strategies were performed for 10 trials, with 30 calibration panels detected per trial. The automated detection and calibration system allowed for these 50 trials and 1500 pattern detections to occur in under 3 hours (using a Pentium IV 3.2 GHz CPU running linux for both image and data processing).

Table 1. Mean Value and percentage of Standard Deviation of the Intrinsic Parameters for each strategy over 10 trials. Deviations are with respect to the mean.

Path	Mean Values				Std. Deviation (% of mean value)			
	f_x	f_y	u_x	u_y	f_x	f_y	u_x	u_y
Stationary	903.2	856.0	233.5	190.6	6.3	5.6	30.9	17.1
2 Panel Translation	785.8	784.3	358.0	206.4	2.7	2.3	3.6	5.0
Rotation	787.7	792.0	324.1	236.6	1.6	1.6	3.9	10.3
Square	781.2	793.1	321.4	274.2	1.2	2.0	2.4	13.9

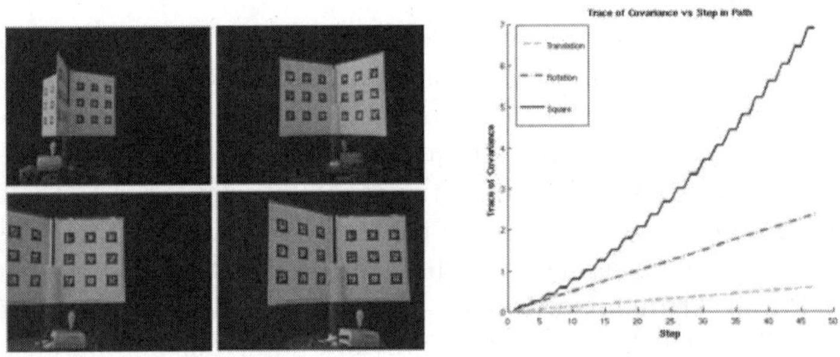

(a) (b)

Fig. 3. (a) Sample Images from Square Pattern. (b) Odometry Error Accumulation for 3 Local Calibration Paths

Table 1 summarizes the intrinsic parameters obtained for each method. The lack of data for the One Panel Translation-only path is due to that, as expected, calibration diverged quite badly in all trials with this method. Other than the stationary method, statistically, the mean parameter estimates are not significantly different between methods.

To examine the difference between odometry buildup among the different paths, each of the three paths which involved motion was simulated. To ensure a fair comparison, path parameters (distance and rotation angles) were scaled accordingly for each trajectory. Fig. 3(b) shows the trace of the covariance matrix as each method progresses. The square pattern accumulates much more odometry error than the other two methods, as expected.

5.2 Mapping an Office Building

To demonstrate the effectiveness of the system when mapping a large space, we instrumented an office environment with 7 camera nodes. The environment

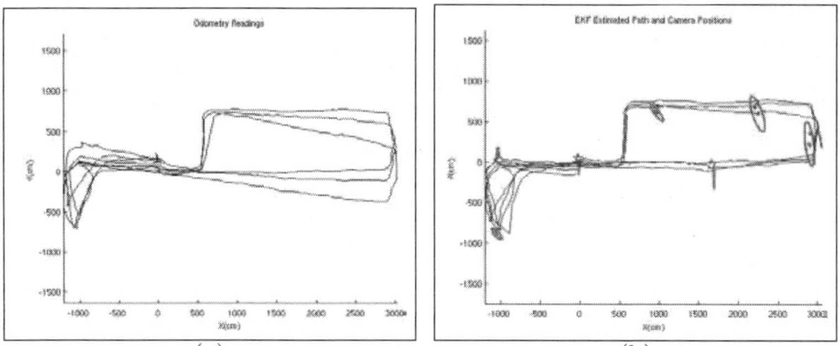

(a) (b)

Fig. 4. (a) Odometry Readings for Hallway Path. (b) EKF Estimate of the Hallway Path. Estimated camera positions with uncertainty ellipses (in red)

consisted of a rectangular loop and a triangular loop connected by a long hallway with length approximately 50 m. The same robot as the previous experiment was used to perform the full calibration and mapping procedure described in Section 3. The robot traversed the environment 3 times and traveled in excess of 360 m in total. The Rotation-only local calibration strategy described in Section 4 was used for simplicity.

From Figs. 4a and b, it is visually clear that the use of camera measurements was able to correct for the buildup of odometry error. However, there are some regions where the filtered path is still a rough approximation since the regions between cameras are traveled without correction of the odometry error. This is most obvious on the far right of the image where there is a very noticeable discontinuity in the filtered path. Since the system does not provide a means for odometry correction between the camera fields of view, this type of behavior is unavoidable without additional sensing.

6 Conclusion

We have outlined an automated method for calibrating and mapping a sensor network of cameras such that the system can be used for accurate robot navigation. The experimental methods show that a system with a very simple level of autonomy can succeed in mapping the environment relatively accurately. A preliminary study was done on local calibration trajectories, which can have a profound effect on the accuracy of the mapping system. Further work in planning and autonomy will likely be the key enhancement in further iterations of this system. The reliance on detection of the calibration target means the robot must move intelligently in order to produce a map of the environment and localize itself within that map.

In this work, we propose the use of a 6-DOF EKF for global mapping. While this approach worked quite well even in a large experiment, it assumes that the system is linear and Gaussian which is a poor approximation in some cases. In particular, the robot builds large odometry errors between cameras and the linearization procedure is only a good approximation when errors are small. A probabilistic method such as Particle Filtering might give improved results in this context, since linearization is not necessary for such a technique.

References

1. *Manual of Photogrammetry.* Am. Soc. of Photogrammetry, 2004.
2. M. Batalin, G. Sukhatme, and M. Mattig. Mobile robot navigation using a sensor network. *International Confernce on Robotics and Automation*, 2003.
3. J. J. Craig. *Introduction to Robotics, Mechanics and Control.* 1986.
4. T.J. Ellis, D. Makris, and J. Black. Learning a multicamera topology. In *IEEE Int. Workshop on Visual Surveillance & Performance Evaluation of Tracking & Surveillance*, pages 165–171, 2003.
5. D. Estrin, D. Culler, K. Pister, and G. Sukatme. Connecting the physical world with pervasive networks. *IEEE Pervasive Computing*, 1(1):59–69, 2002.
6. O. D. Faugeras. *Three-Dimensional Computer Vision.* MIT Press, 1993.
7. M. Fiala. Artag revision 1, a fiducial marker system using digital techniques. In *National Research Council Publication 47419/ERB-1117*, November 2004.
8. M. Fiala. Vision guided robots. In *Proc. of CRV'04 (Canadian Confernence on Computer and Robot Vision*, pages 241–246, May 2004.
9. W. E. L. Grimson, C. Stauer, R. Romano, and L. Lee. Using adaptive tracking to classify and monitor activities in a site. In *Proc. of the IEEE Conf. on Computer Vision and Pattern Recognition*, pages 22–29, 1998.
10. Andrew Howard, Maja J Mataric, and Gaurav S. Sukhatme. Localization for mobile robot teams using maximum likelihood estimation. In *Proceedings of the IEEE/RSJ International Conference on Intelligent Robots and Systems*, pages 434–459, EPFL Switzerland, 2002.
11. R. Kurazume and S. Hirose. An experimental study of a cooperative positioning system. *Autonomous Robots*, 8(1):43–52, Jan. 2000.
12. J.J. Leonard and H.F. Durrant-Whyte. Mobile robot localization by tracking geometric beacons. *IEEE Trans. on Robotics & Automation*, 7(3):376–382, 1991.
13. Dimitris Marinakis, Gregory Dudek, and David Fleet. Learning sensor network topology through monte carlo expectation maximization. In *Proc. of the IEEE International Conference on Robotics & Automation*, Spain, 2005.
14. I. Poupyrev, H. Kato, and M. Billinghurst. Artoolkit user manual, version 2.33. *Human Interface Technology Lab, University of Washington*, 2000.
15. Ioannis M. Rekleitis, Gregory Dudek, and Evangelos E. Milios. On the positional uncertainty of multi-robot cooperative localization. Multi-Robot Systems Workshop, Naval Research Laboratory, Washington, DC, USA, March 18-20 2002.
16. I. Rekletis and G. Dudek. Automated calibration of a camera sensor network. In *IEEE/RSJ Int. Conf. on Intelligent Robots & Systems*, pages 401–406, 2005.
17. Stergios I. Roumeliotis and George A. Bekey. Distributed multirobot localization. *IEEE Transactions on Robotics and Automation*, 18(5):781–795, 2002.
18. R. Smith, M. Self, and P. Cheeseman. Estimating uncertain spatial relationships in robotics. *Autonomous Robot Vehicles*, pages 167 – 193, 1990.
19. R. Y. Tsai. Synopsis of recent progress on camera calibration for 3-d machine vision. *The Robotics Review*, pages 147–159, 1989.
20. R. Y. Tsai and R. K. Lenz. A versatile camera calibration technique for high-accuracy 3d machine vision metrology using off-the-shelf tv cameras and lenses. *IEEE Journal of Robotics and Automation*, pages 323–344, 1987.
21. R. Y. Tsai and R. K. Lenz. Real time versatile robotics hand/eye calibration using 3d machine vision. *IEEE Int. Conf. on Robotics & Automation*, 1988.
22. Z. Zhang. A flexible new technique for camera calibration. *IEEE Transactions on Pattern Analysis and Machine Intelligence*, 22(11):1330–1334, 2000.

Adaptive Robotic Communication Using Coordination Costs*

Avi Rosenfeld, Gal A Kaminka, and Sarit Kraus

Bar Ilan University, Ramat Gan, Israel
{rosenfa, galk, sarit}@cs.biu.ac.il

Summary. Designers of robotic groups are faced with the formidable task of creating effective coordination architectures that can deal with changing environment conditions and hardware failures. Communication between robots is one mechanism that can at times be helpful in such systems, but can also create a time and energy overhead that reduces performance. In dealing with this issue, various communication schemes have been proposed ranging from centralized and localized algorithms, to non-communicative methods. In this paper we argue that using a coordination cost measure can be useful for selecting the appropriate level of communication within such groups. We show that this measure can be used to create adaptive communication methods that switch between various communication schemes. In extensive experiments in the foraging domain, multi-robot teams that used these adaptive methods were able to significantly increase their productivity, compared to teams that used only one type of communication scheme.

1 Introduction

Groups of robots are likely to accomplish certain tasks more quickly and robustly than single robots [3, 5, 7]. Many robotic domains such as robotic search and rescue, demining, vacuuming, and waste cleanup are characterized by limited operating spaces where robots are likely to collide. In order to maintain group cohesion under such conditions, some type of information transfer is likely to be helpful in facilitating coherent behavior in robotic group tasks and thus better achieve their task. This is especially true as robotic domains are typically fraught with dynamics and uncertainty such as hardware failures, changing environmental conditions, and noisy sensors.

Questions such as what to communicate and to whom have been the subject of recent study [7, 11, 12]. In theory, communication should always be advantageous–the more information a robot has, the better. However, assuming communication has a cost, one must also consider the resources consumed in communication, and whether the cost of communication appropriately matches the needs of the domain. We believe that different communication schemes are best suited for different environmental conditions. Because no one communication method is always most effective, one

* This work was partially supported by the Israeli Ministry of Science and Technology.

way to improve the use of communications in coordination is to find a mechanism for switching between different communication protocols so as to match the given environment.

This paper provides such an adaptive communications framework using a coordination cost measure that quantifies all resources spent on coordination activities. Our model explicitly includes resources such as the time and energy spent communicating. In situations where conflicts between group members are common, more robust means of communication, such as centralized models, are most effective. When collisions are rare, coordination methods that do not communicate and thus have the lowest overhead, work best.

We present two novel domain-independent adaptive communication methods that use communication cost estimates to alter their communication approach based on domain conditions. In our first approach, robots uniformly switch their communication scheme between differing communication approaches. In this method, robots contain full implementations of several communication methods, and switch between them as needed. In contrast, our second approach represents a generalized communication scheme, that allows each robot to adapt independently to its domain conditions. Each robot creates its own communication range radius (which we refer to as its *neighborhood of communication*), to create a sliding scale of communication between localized to centralized methods. Each robot uses its coordination cost estimate to determine how large its neighborhood should be.

To evaluate these adaptive methods, we performed thousands of trials using an established robotic simulator, in a multi-robot foraging task. We tested groups of varying sizes and communication methods. We found that groups that used the adaptive methods often significantly exceeded the best productivity levels of the non-adaptive algorithms they were based on.

2 Related Work

A major challenge to designers of robotic groups exists in choosing an optimal communication method. Many practical frameworks have been presented for use within robotic teams [3–9, 12] and can generally be assigned to categories of no communication, localized, and centralized approaches.

It is possible to create effective group behavior without any communication [2]. For example, the Stigmergy concept [6] involves group members basing their actions by observing how other group members previously modified their environment. This approach has been shown to be effective in several animal and robotic domains [6] without using any explicit communication. Coordination without communication can potentially facilitate better adaptability, robustness and scalability qualities over methods using communication [11]. Additionally, the lack of communication also allows such methods to be implemented on simpler robots. However, such algorithms often require powerful and accurate sensing capabilities [9]. Also, our results demonstrate that groups implementing these methods did not always provide the highest levels of productivity, especially within dynamic domains where frequent coordination conflicts exist.

A second set of approaches attempt to improve group performance by having robots locally communicate information [7, 9]. For example, work of Jäger and Nebel [7] present a method whereby robots nearing a collision stopped to exchange trajectory information. They then successfully detect and resolve deadlock conditions of two or more robots mutually blocking. However, their trajectory planning method was not able to perform well in groups of over five robots. In contrast, Mataric [9] reported that a local communication scheme scaled well with group size. One key difference seems to lie within the local communication implementations. In Jäger's algorithm, one or more robots must stop moving during trajectory replanning. We believe this led to a breakdown in the system once the group size grew. Mataric's locally communicating robots broadcast information while continuing their foraging task. This allowed for better scalability qualities.

A third type of approach involves the use of some type of central repository of information [12]. This centralized body, which could also be implemented as one "expert" teammate, would then be able to easily share its store of pooled information with other teammates. While this approach allows for free information sharing and can thus improve performance, several drawbacks are evident. First, the centralized mechanism creates a single point of failure. The cost of communication is also likely to be large, and requires hardware and bandwidth suitable for simultaneous communication with the centralized body. While these drawbacks are at times significant, they may be justified given the needs of the domain.

In this work, we assume that representative communication methods from these categories are predefined, and have been implemented with optimal values for their exact parameters given domain conditions. Several approaches exist for finding these parameters within a given coordination method. For example, work by Yoshida et al. [4] presented a framework to derive an optimal localized communication area betweens within groups of robots to share information in a minimum of time. This approach assumes domain conditions such as spatial distributions and the probability of information transmission can be readily calculated. Previously, Goldberg and Mataric [5] focused on *interference* (which they defined as the time robots spent colliding) as a basis for measuring a coordination method's effectiveness. However, they did not address how to create adaptive methods based on interference. Our previous work [10] built upon this interference definition to include all resources spent resolving coordination conflicts including the time spent before and after collisions. We then demonstrated that parameter tweaking is possible through this measure. The advantage to this approach over the work of Yoshida et al. [4] is its ability to allow robots to autonomously adapt, even in dynamic environments. However, in contrast to their work, our previous work [10] did not study communication issues.

In this work, we use coordination cost measures to compare a given set of communication methods and to create adaptive methods based on these existing methods. We explicitly model all resources spent on coordination activities including the resources spent on communication even if they do not detract from the time to complete the task. Our goal was to properly match communication methods to domain conditions, while considering their relative costs. Furthermore, adaptation between communication schemes presents new challenges, since many protocols require standardized communication between all team members. These challenges are addressed in this paper.

3 Using Coordination Costs to Adapt Communications

Our coordination cost measure facilitates identifying which communication method is most suitable given the environment. We model every robot's coordination cost C_i, as a factor that impacts the entire group's productivity. We analyze two cost categories: (i) costs relating to communication and (ii) proactive and/or reactive collision resolution behaviors. We focus on the time and energy spent communicating and in the consequent resolutions behaviors (see Implementation Section for full details). We then combine these factors to create a multi-attribute cost function based on the Simple Additive Weighting (SAW) method [14] often used for multi-attribute utility functions. While methods with no communication have no C_i for communication, this method could not always successfully resolve collisions and then spent more resources on collision resolution behaviors, or another C_i. Conversely, centralized methods incurred a communication cost C_i that often eclipsed the needs of the domain and weighed heavily on productivity. Other communication issues, such as bandwidth limitations, can similarly be categorized as additional cost factors as they impact any specific robot. For example, if a robot needed to retransmit a message due to limited shared bandwidth, costs in terms of additional time latency and energy used in retransmission are likely to result.

Using this measure we can compare communication methods, but in this paper we focus on using it for online adaptation between communication schemes. In this section we present two types of adaptive methods: (i) uniform communication adaptation (ii) adaptive neighborhoods of communication. Both methods led to significant increases in productivity over static approaches (see Experiments section).

3.1 Uniform Switching Between Methods

In our first method, all robots simultaneously switch between mutually exclusive communication methods as needed. In order to facilitate this form of adaptation, each robot autonomously maintains a cost estimate, V used to decide which communication method to use. As a robot detects no resource conflicts, it decreases an estimate of this cost, V, by an amount W_{down}. When a robot senses a conflict is occurring, the value of V is increased by an amount W_{up}. The values for V are then mapped to a set of communication schemes methods ranging from those with little cost overhead such as those with no communication, to more robust methods with higher overheads such as the localized and centralized methods. As the level of projected conflicts rises (as becomes more likely in larger group sizes) the value of V rises in turn, and the robots use progressively more aggressive communication methods to more effectively resolve projected collisions. While these activities themselves constitute a cost that detracts from the group's productivity, they are necessary as more simple behaviors did not suffice. As different coordination methods often have different costs, C_i for a given domain, we believed this approach could be used to significantly improve the productivity of the group.

Several key issues needed to be addressed in implementing this method with groups of robots. First, we assumed that all group members are aware of the overheads associated with various coordination methods, and can order them based on their relative complexities. This ordering can be derived from theoretical analysis or through observation (as we do in later in this paper). Second, an approach to quickly set the weights, W_{up}, and W_{down} used within our algorithms is needed. While traditional learning methods, such as Q-learning [13] may converge on an optimal policy, this approach is difficult to implement because of two major reasons. First, Q-learning is based on a a concept of "state" that is not readily definable during task execution. As opposed to clearly defined discrete domains, there is no reward for any given cycle of activity in the robotic domains we studied. Even assuming an optimal policy could be learned, a second, more fundamental problem exists. Robotic domains often contain dynamics that render a learned policy obsolete very quickly. Thus, our approach is to sacrifice finding a globally optimal policy in exchange for finding a locally optimal policy after a much shorter training period for our weights. We used a gradient learning procedure to achieve this result.

Next, it must be noted that uniform adaptation requires all robots to change communication in sync because of the mutual exclusivity of the methods used. For example, it is impossible for one robot to use a centralized method, with others using one without communication, as the centralized approach is based on information from all team members. As a result, once any one robot in the group autonomously decided it needed to switch communication schemes, a communication change must also occur within all other team members. This could force certain members to use a more expensive communication method than it locally found necessary. We relaxed these requirements in the second adaptive method, presented in the next section.

Finally, care must be taken to prevent the robots from quickly oscillating between methods based on their localized conditions. In our implementation, communication adaptation was triggered once one robot's value for V exceeded a certain threshold. After this point, that robot broadcasted which method it was switching to and all group members would change in kind and reinitialize their cost estimates V to this new value. Furthermore, we also used domain specific information, such as prioritizing collisions closer to the home base within our foraging domain. In this fashion, we partially limited the types of triggers to those of importance to the entire group. Once again, our second type of communication adaptation relaxes this requirement and is effective without any such heuristics.

3.2 Adaptive Neighborhoods of Communication

The advantage in our first adaptive approach lies in its simplicity. Our uniform adaptive approach switches between existing coordination methods based on estimated coordination cost. Assuming one analyzes a new domain with completely different communication methods, and can order the communication methods based on their communication costs, this approach will be equally valid as it implements existing methods and reaches the highest levels of productivity from among those methods–whatever they may be.

In contrast, our second adaptation method is a parameterized generalization of the three specific categories of communication methods (No-Communication, Localized, and Centralized). As many robotic domains use elements of these same methods [3, 4, 6–8, 12], we reason that a similar approach is likely to work in these and other domains as well.

The basis of this approach is introducing a parameter to control how large a radius of communication is used by each robot. This method uses a distance d inside which robots exchange information, which we term its communication neighborhood. Formally, this radius of communication could be considered a neighborhood Γ of size d, created from robot v and includes all teammates, u, inside this radius. As such, we represent the neighborhood as $\Gamma_d(v) = \{u|\ u \text{ robot}, dist(u,v) \leq d\}$.

Adjusting the value of d in Γ_d can be used to approximate the previously studied communication categories. Assuming d is set to zero, no communication will ever be exchanged and this method is trivially equivalent to the No-Communication method. Assuming d is set to some small amount, ε, this method will become similar to the Localized method and information will be exchanged only with the robot it is about to collide with. If d is set to the radius of the domain, the neighborhood of communication encompasses all teammates this method becomes similar to the Centralized method. Thus, the degree of centralization exclusively depends on the value of d.

4 Implementation Details

We used the Teambots [1] simulator to implement communication schemes involving no communication, localized and centralized approaches within groups of Nomad N150 foraging robots. The foraging domain is defined as locating target items from a search region *S*, and delivering them to a goal region *G* [5]. Foraging robots often collide as they approach the home base(s) within their area of operation. In our domain there was only one goal region, *G*, which was located in the center of the operating area. In our implementation, there were a total of 60 target pucks spread throughout an operating area of approximately 10 by 10 meters. We measured how many pucks were delivered to the goal region within 9 minutes by groups of 2–30 robots using each communication type. We averaged the results of 100 trials for each group size with the robots being placed at random initial positions for each run. The number of trials performed and the relatively large group sizes would have been difficult to implement on physical robots.

We created experiment sets measuring the time and energy spent in two coordination categories–communication and collision resolution. The coordination costs in our first set of experiments involved the time spent in communication and collision resolution behaviors out of each trial's total time of 9 minutes. In our second set of experiments, we allocated each robot 500 units of fuel. We assumed most of the fuel was used by the robots to move, with a smaller amount (1 unit per 100 seconds) used to maintain basic sensors and processing. For the time based experiments, we assumed robots pairs stopped for 1/5 of a second to communicate, representing some methods [7] where robots stop to exchange information. In the energy based localized experiments, we assumed robots did not stop to communicate, as is the case with other methods [9], but each robot still spent 0.3 units of fuel per communication exchange. Our coordination cost involved the amount of fuel that was used in communication and repulsion behaviors.

All three communication schemes were similar in that they resolved collisions by mutually repelling once they sensed a teammate within a certain safe distance ε, which we set to 1.5 robot radii. Once within this distance, robots acted as they were in danger of colliding and used repulsions schemes to resolve their collision(s). The No-Communication method was unique in that robots never used time or fuel to communicate, and thus only had costs relating to the repulsion behaviors robots engaged in. However, this method assumed domain specific information, namely it used the robot's autonomously computed scalar distance, S, from its location to the home base in the domain. Robots used a function of this distance, which we implemented to be 5 times S and rounded to the closest second, as the time to repel from its teammate(s) after a projected collision.

Our localized method used less domain specific information and is similar to the localized methods previously proposed [7, 9]. Communication between robots was initiated once it was in danger of colliding–a teammate came within the ε distance. After this event, these group members would exchange information above their trajectories (here their relative distances from their typical target, their home base). The closer robot then moved forward, while the other robot repelled for a fixed period of 20 seconds.

Our final method, *Centralized*, used a centralized server with a database of the location of all the robots similar to other centralized methods [12]. Within this method, one of two events triggered communication. First, as with the localized method, robots dropping within the ε distance initiated communication by reporting its position, done here with the centralized server. The server then reported back a repel value based on its relative position to all other teammates. However, in order for the server to store a good estimate of the positions of all robots, a second, often more frequent type of communication was needed where each robot reported its position to the server with frequency L. If this communication occurred too frequently, this central database would have the best estimate of positions, but the time or energy spent on communication would spike, and productivity would plummet. If communication was infrequent, the latency of the information stored on the server would create outdated data. This in turn would reduce the effectiveness of this method, and result in more collisions. We found that a latency time of 1 second yielded the highest productivity in the time based experiments, and a latency value of 5 seconds yielding the highest productivity within the energy based experiments.

It is important to stress that the focus of our work is switching between categories of communication methods, and not to find optimal parameters within any one given communication method. We refer the reader to previous work [4, 10] on how to theoretically or empirically derive parameters within one communication method. Our work is based on the understanding that a high negative correlation exists between each groups' productivity and our coordination cost, regardless of the exact implementation for the parameters used in the No-Communication, Localized and Centralized methods. For example, we studied 7 latency variations within the Centralized method in both experiment sets. Our groups enforced maximal latency periods of L set to 0.1, 0.3, 1, 5, 10, 30 and 60 seconds. While the optimal latency value was different across experiment sets, in both cases the productivity of these variations was highly negatively correlated with their relative coordination costs. In the first case, we found a correlation of -0.95 between these latency variations and the corresponding coordination cost based on time. In the trials based on fuel, this value was -0.97. Similarly, factors such as the exact energy spent on communication exchanges (0.3 units of fuel per exchange) or the time spent on communication (1/5 seconds per exchange) could vary across domains, or the distance between communication partners. However, we consistently found that the resources spent on these communication exchanges was strongly negatively correlated with those groups' productivity.

While we consider the neighborhood communication approach to be a parameterized generalization of the three previously described categories, some implementation details differ in this method over the static ones it emulates. Within this method, once any robot A, detects another robot within the ε distance, it initiates communication with all robots found within the $\Gamma_d(A)$ area. All robots in $\Gamma_d(A)$ must then report back to Robot A with their projected trajectories. Robot A then sorts all robots' trajectories by their relative distances from the home base in the domain. This robot then reports back to every robot within $\Gamma_d(A)$ a repel value based on that robot's relative position in the neighborhood. All robots, including the initiating robot (robot A), then accept this value and immediately engage in repel behaviors for the dictated length of time. It is possible that a robot may be a member of more than one neighborhood. In such cases, robots accept the larger repel value regardless of the sender.

While the repel amounts of the robot initiating communication (Robot A) are calculated in a similar fashion to the previously described centralized method, here these values are calculated by members of the team, instead of one centralized server. The radius of communication in the centralized approach is the full width of the domain, while the Γ_d radius is typically much smaller. However, the biggest difference in implementing this approach is how repel values are obtained. Robots in previous methods only repelled based on communication received after dropping within the ε distance. In this method, robots may repel if they enter the Γ_d radius even if they are not in immediate danger of colliding. The reason for this is as follows. As robots within the Γ_d radius are typically close to each other, we found that these robots often would soon initiate their own radii of communication. In other methods this was not a concern, as other teammates were not effected by this phenomenon. However, here this would create multiple neighborhoods involving the same teammates. Thus, proactively assigning repel values was crucial for containing communication costs as Γ_d grew.

5 Experimental Results

The first set of experiments attempts to first lend support to the underlying hypothesis, that the combined coordination cost measure is in fact correlated to the productivity of the different groups. Our results from experiments involving time and energy costs support the claim that the best method of communication does change with domain conditions (see figure 1). In the time experiments, we found an average correlation of -0.96 between the average productivity found in groups of 2–30 robots and the group's corresponding average cost. In the equivalent energy based experiments, we found a value of -0.95.

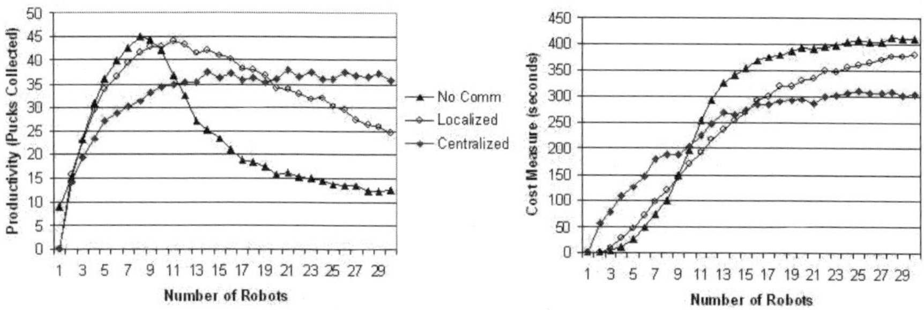

Fig. 1. Comparing the productivity levels of three communication types with the coordination costs based on the time spent on communication relative to different group sizes. Results averaged from 100 trials per datapoint.

Similarly, we found that no one neighborhood size always fared best. We compared the productivity levels of foraging groups where d was set to 1, 2, 3, 5 and 50 robot lengths. Recall that ε is approximately 1 robot length (1.5 radii). Thus Γ_1 represents the nearly localized variation with Γ_{50} corresponding to the nearly centralized version of this method.

Figure 2 represents the relative productivity levels for these static neighborhood groups relative to the energy costs levels measured in these groups. Notice how in small groups, Γ_1 yielded the highest average productivity. As we have seen, when possible, resources spent on coordination, here by creating large communication neighborhoods, should be avoided when possible. As small areas of communication sufficed in small groups, this approach had the highest productivity. As the group size grew, additional communication was necessary to maintain high productivity levels. As a result, larger neighborhoods were necessary and groups with Γ_5 resulted in the highest productivity. However, forcing too much communication when not necessary created communication costs that reduced productivity to levels found in methods that spend too few resources on communication. In this method, the productivity level of the Γ_{50} method, which created too large a neighborhood, approached those of Γ_1, which did not create a large enough one. We again found a strong correlation between the various Γ_d variations and the groups' corresponding coordination costs and productivity with an average negative correlation of -0.96.

Based on the confirmed hypothesis, that the cost measure is indeed correlated (negatively) with performance, the next set of experiments evaluated the performance of the two adaptive methods compared to the static methods on which they were based. Figure 3 shows the results from these experiments. Notice that both adaptive approaches approximated or significantly exceeded the highest productivity levels

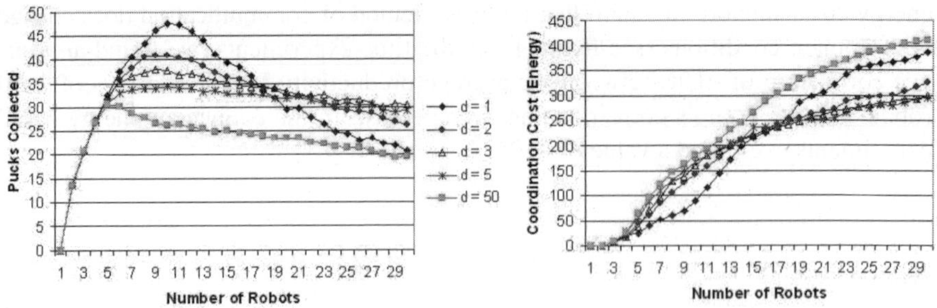

Fig. 2. The impact of varying neighborhood sizes (d) on productivity levels and costs in energy experiments. Results averaged from 100 trials per datapoint.

of the static methods (No Communication, Local, and Centralized methods) they were based on, especially in medium to large groups. We attribute the success of both methods to their ability to change communication methods to the needs of the domain. We believe that the neighborhood method outperformed the uniform one as it was allowed to create locally different neighborhood sizes, something none of the static neighborhood methods were capable of. This in turn facilitated better adaptation and higher productivity.

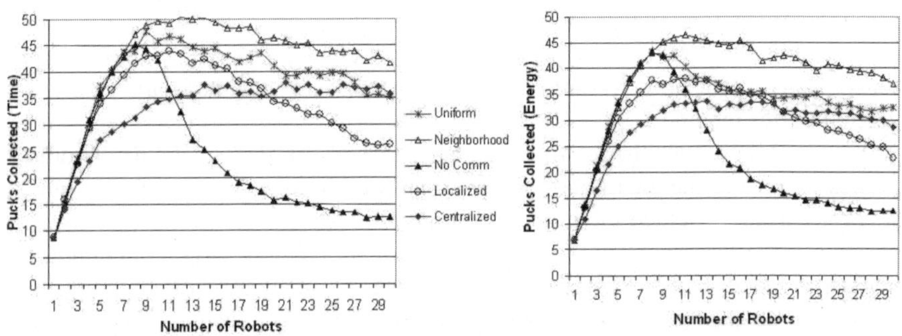

Fig. 3. Comparing adaptive communication methods based on time and energy costs to static methods. Results averaged from 100 trials per datapoint.

To evaluate the statistical significance of these results, we conducted the two tailed t-test and a 1-factor ANOVA test comparing our adaptive groups and the three static groups they were based on. In all cases, in both time and energy categories, the null hypothesis p values were below 0.001. This confirms our hypothesis that we can improve productivity through creating adaptive methods based on communication costs.

6 Conclusion

This work demonstrates how coordination costs can account for the relative effectiveness of robotic communication methods. Our measure focuses on the time and energy spent communicating and resolving collisions. We demonstrate the effectiveness of our methods in comparing between very different communication methods falling within categories of no communication, localized and centralized communication methods. By using this information we are able to match the most effective communication scheme to a given robotic domain. We present two general adaptive communication algorithms, uniform and neighborhood methods. We show, in thousands of foraging experiments, that coordination cost is indeed negatively correlated with productivity, and that the use of our adaptive methods leads to significant performance boosts. While we find the neighborhood adaptive method to be more effective in the robotic foraging domain we studied, both approaches are likely to be applicable to many other domains [3, 7, 8, 12]. It is possible that the uniform method is easier to implement or will yield better adaptive qualities in other domains.

References

1. T. Balch. www.teambots.org, 2000.
2. T. Balch and R. Arkin. Behavior-based formation control for multirobot teams. *IEEE Transactions on Robotics and Automation*, 14(6):926–939, December 1998.
3. G. Dudek, M. Jenkin, and E. Milios. A taxonomy for multi-agent robotics. *Robot Teams: From Diversity to Polymorphism, Balch, T. and Parker, L.E., eds., Natick, MA: A K Peters*, 3:3–22, 2002.
4. M. Y. J. O. Eiichi Yoshida, Tamio Arai. Local communication of multiple mobile robots: Design of optimal communication area for cooperative tasks. *Journal of Robotic Systems*, 15(7):407–427, 1998.
5. D. Goldberg and M. Matarić. Design and evaluation of robust behavior-based controllers for distributed multi-robot collection tasks. In *Robot Teams: From Diversity to Polymorphism*, pages 315–344, 2001.
6. O. Holland and C. Melhuish. Stigmergy, self-organization, and sorting in collective robotics. *Artif. Life*, 5(2):173–202, 1999.
7. M. Jager and B. Nebel. Decentralized collision avoidance, deadlock detection, and deadlock resolution for multiple mobile robots. In *IROS*, pages 1213–1219, 2001.
8. G. A. Kaminka and R. Glick. Towards robust multi-robot formations. In *ICRA-06*, 2006.
9. M. J. Matarić. Using communication to reduce locality in multi-robot learning. In *AAAI/IAAI*, pages 643–648, 1997.
10. A. Rosenfeld, G. Kaminka, and S. Kraus. Adaptive robot coordination using interference metrics. In *The Sixteenth European Conference on Artificial Intelligence*, pages 910–916, August 2004.
11. S. Sen, M. Sekaran, and J. Hale. Learning to coordinate without sharing information. In *Proceedings of the Twelfth National Conference on Artificial Intelligence*, pages 426–431, Seattle, WA, 1994.
12. A. Tews. Adaptive multi-robot coordination for highly dynamic environments. In *CIMCA*, 2001.
13. C. J. C. H. Watkins. Learning from delayed rewards. *Ph.D. Dissertation, Kings College*, 1989.
14. K. Yoon and C. Hwang. *Multiple attribute decision making: an introduction.* Prentice Hall, Thousand Oaks: Sage, 1995.

What to Communicate? Execution-Time Decision in Multi-agent POMDPs

Maayan Roth[1], Reid Simmons[2], and Manuela Veloso[3]

[1] Robotics Institute, Carnegie Mellon University mroth@andrew.cmu.edu
[2] Robotics Institute, Carnegie Mellon University reids@cs.cmu.edu
[3] Computer Science Department, Carnegie Mellon University veloso@cs.cmu.edu

Summary. In recent years, multi-agent Partially Observable Markov Decision Processes (POMDP) have emerged as a popular decision-theoretic framework for modeling and generating policies for the control of multi-agent teams. Teams controlled by multi-agent POMDPs can use communication to share observations and coordinate. Therefore, policies are needed to enable these teams to reason about communication. Previous work on generating communication policies for multi-agent POMDPs has focused on the question of **when** to communicate. In this paper, we address the question of **what** to communicate. We describe two paradigms for representing limitations on communication and present an algorithm that enables multi-agent teams to make execution-time decisions on how to effectively utilize available communication resources.

1 Introduction

The problem of generating optimal policies for multi-agent POMDPs is known to be NEXP-complete [1], making optimal policy-generation intractable. Therefore, the bulk of recent work in this area has focused on finding heuristic algorithms that can generate high-quality policies for multi-agent teams in a reasonable amount of time. Team members can improve their ability and the abilities of their teammates to reason about their environment by communicating their local observations We are interested in studying heuristics for making execution-time communication decisions [5]. By assuming free communication at policy-generation time, we can generate centralized policies for multi-agent teams using a single-agent POMDP solver. We then reason about communication at execution time to enable decentralized execution of these policies.

In situations where teams have the capability to perform free and unlimited communication, the best strategy is for each agent to broadcast all of its observations to its teammates [4]. However, in general, communication is not free. To use communication effectively, multi-agent teams must trade off the benefit that can be achieved through communication with the cost of communicating. We consider two communication paradigms:

Fixed cost per communication instance - Every time the agents decide to communicate, they incur a known fixed cost [4, 6, 7]. The question that must be answered by a communication heuristic is, in essence, **when** to communicate. In some previous approaches, the use of this paradigm to focus on the question of when to communicate has justified algorithms in which agents are required to communicate their entire observation histories if they determine that communication is beneficial [8, 5]. In this paper, we extend this paradigm to include the case in which cost scales with the amount of information to be communicated. We model this by assuming a fixed cost per observation transmitted.

Limited communication bandwidth - Often, communication among agents has a strict limit on available bandwidth. For example, in robot soccer, an attempt to communicate the complete and frequent sensory data would easily overload the available communication resources [11]. In planetary exploration, communication bandwidth is limited, and in a proposed communication architecture, agents need to share a limited number of communication relays in order to stay in contact with their teammates [12]. Other domains include distributed surveillance, in which the observations themselves are very large [13]. In general, it is possible to quantify the amount of bandwidth available for communication between teammate agents. The challenge, then, is to determine **what** to communicate so as to best use the available bandwidth, an issue that has received little attention in the current multi-agent POMDP literature.

Our previous work presented an algorithm for making execution-time decisions about **when** to communicate that allows agents to successfully execute centralized policies in a decentralized fashion [5]. In this paper, we introduce an algorithm that builds on our previous work to address the question of **what** to communicate. We show the applicability of this algorithm to both communication paradigms presented above, and verify the success of our algorithm through experimental results. When there is a fixed communication cost per observation, our algorithm allows agents to identify only those observations that are relevant to team performance. In the case of bandwidth limitation, where each agent is allowed to communicate a fixed number of observations per unit time, our algorithm enables agents to choose those observations that will most improve expected team reward.

Given the complexity of generating policies for multi-agent POMDPs, it is not surprising that approaches have so far been validated on very small test problems. The *multi-agent tiger domain* , introduced by Nair *et al.* [2], has emerged as a commonly-used benchmark case [9, 5, 10]. It has the significant advantage of being small enough for easy use in explanatory examples, while still containing a challenging coordination problem. In this paper, we introduce a new domain, called the *Colorado/Wyoming problem.* This new domain contributes several key attributes that make it useful for evaluating communication heuristics, such as the presence of multiple different observations that provide varying qualities of information. Together with the multi-agent tiger domain, it is a step toward the compilation of a comprehensive suite of benchmark domains for multi-agent POMDPs.

2 Multi-agent POMDPs

Several representations (e.g. DEC-POMDP [1], MTDP [4], POIPSG [3], IPOMDP [10]) can be used to model cooperative teams of agents operating under partial observability. In this paper, we use the notation introduced in [1], which defines a DEC-POMDP as a tuple $\langle \alpha, \mathcal{S}, \mathcal{A}, \mathcal{T}, \Omega, \mathcal{O}, \mathcal{R} \rangle$ where α is the number of agents in the team, $\mathcal{S}$ is the set of n world states, and $\mathcal{A}$ is the set of m possible joint actions of the team, where each joint action, a^i, is composed of α individual actions. $\mathcal{T}$, the transition function, depends on joint actions and gives the probability associated with starting in a particular state s^i and ending in a state s^j after the team has executed the joint action a^k. Ω is the set of possible joint observations, where each joint observation, ω^i, is composed of α individual observations. The observation function, $\mathcal{O}$, gives the probability of observing the joint observation ω^i after taking action a^k and ending in state s^j. $\mathcal{R}$ indicates the reward that is received when the team starts in a state s^i and takes the joint action a^k. In this paper, we present example domains in which the agents are identical. However, this is not a necessary property of multi-agent POMDPs and our algorithms are equally applicable to heterogeneous teams.

It is important to note that, although the observation function is given in terms of joint observations, each agent observes only its own individual observations. Additionally, when executing a policy, the individual agents receive no explicit notification of the actions that were taken by their teammates. Multi-agent POMDPs are challenging to solve because, to accurately model the state and choose a policy, each agent must reason not only over uncertainty in the environment, but also about the possible behaviors of its teammates.

There are several classes of observability possible in cooperative multi-agent teams. In this paper, we examine domains with *collective partial observability*, meaning that even if every agent on the team had access to the local observations of all of its teammates, this union of team information may still be insufficient to uniquely identify the world state. The algorithms that we discuss in this work are equally applicable to domains with *collective observability*, sometimes called DEC-MDP, in which the union of individual observations is sufficient to uniquely determine the team's state.

3 DEC-COMM: Deciding When to Communicate

The problem of generating optimal policies for multi-agent POMDPs is known to be NEXP-complete [1], making exact solutions unfeasible and necessitating the use of heuristics. Our previous work [5] developed an approach that exploits a known property of multi-agent POMDPs, namely that the presence of free and unrestricted communication can be used to transform a multi-agent POMDP into a centralized, or single-agent, POMDP [4], a problem that has a smaller complexity of PSPACE [14]. The approach of our previous work is to assume, at policy-generation time, that communication is free, allowing agents to know the local observations of their teammates at every timestep. This enables us to write the multi-agent POMDP as a single-agent POMDP. We are then able to use any single-agent POMDP solver (e.g. [15]) to generate a centralized policy for the team.

The challenge, then, is to enable agents to execute the centralized policy in a decentralized manner despite the fact that, in general, communication is not free. Because the transition and observation functions of a multi-agent POMDP depend on the joint action, an individual member of the team cannot compute belief independently. To correctly execute a centralized policy, the agents must form the same approximation of joint belief, and each agent must ensure that it is selecting the same joint action as its teammates. This requires agents to model joint belief based only on information that is globally available to all of the teammates.

Our approach is to have each agent calculate $\mathcal{L}^t$, the distribution of possible joint beliefs of the team. Each element, $\mathcal{L}_i^t$, is a possible joint observation history. $\mathcal{L}_i^t$ is defined as the tuple $\langle b^t, p^t, \boldsymbol{\omega}^t \rangle$, where $\boldsymbol{\omega}^t$ is the joint observation history leading to $\mathcal{L}_i^t$, b^t is the joint belief given that history, and p^t is the probability of the team observing that history. [5] provides a detailed algorithm, GROWTREE, that describes how this tree is calculated. The important detail is that the contents of this tree do not depend on any agent's local observations. Therefore, all the agents can compute identical trees independently.

Agents can calculate a joint action over the distribution of possible joint beliefs and be assured that the joint action selected is identical across teammates. This action selection can by done by means of any function that operates over the leaves in $\mathcal{L}^t$. We introduced one possible EVALUATE function, Q-POMDP, in [5]. However, since agents do not use their local observations to refine their beliefs about the state of the world, the selected action is unaffected by the agents' true experiences. Communication provides a means for agents to share their local observations with their teammates, enabling the team to use those observations when making decisions.

Since communication is not free, we want to reason about when to communicate. The DEC-COMM algorithm, presented in detail in [5], enables agents to make execution-time decisions about when to communicate their observations to teammates. An agent can hypothesize about the joint action that would be selected by the team if it chose to communicate by pruning $\mathcal{L}^t$ of all of the observation histories that are inconsistent with its own local observations. It compares the expected reward of this new joint action, a_C, with the expected reward of the joint action that would be chosen if it does not communicate, a_{NC}. If the change in expected reward is above some threshold ϵ (the cost of communication), the agent broadcasts its observation history to its teammates, who then prune their own $\mathcal{L}^t$ to be consistent with the communicated observations.

4 Choosing What to Communicate

The DEC-COMM algorithm described above answers the question of **when** to communicate, making run-time communication decisions in the context of decentralized execution of a centralized policy. However, it does not address the question of **what** to communicate. Each time an agent communicates, the algorithm requires it to broadcast all of the observations received since the last time that it communicated. There are several shortcomings to this approach. First, it is unnecessarily wasteful, forcing agents to broadcast observations that do not serve to improve team performance. Second, it can deal only with communication limitations that are represented through a fixed cost of communication. The algorithm in its original form does not enable agents to make efficient use of limited communication bandwidth.

Given a limited bandwidth availability of k observations, the goal is to find those k observations that would most increase the expected team reward if communicated. While this can be done exhaustively by calculating the value of information of each subset of size k of an agent's observation history, it is intractable for run-time decision making.

Instead, we introduce the BUILDMESSAGE heuristic. The intuition is as follows: The agent can calculate a_C, the joint action that the team would perform if the agent could broadcast its entire observation history. From this agent's perspective, a_C is the best possible action that the team could take, given all of the available information. If a_C is the same as the action that would be performed without communication, it is clear that the agent cannot expect that communicating only a subset of observations will improve expected reward. If, however, communication could potentially improve the team's selection of a joint action, then it seems logical to select those observations that most increase the desirability of choosing a_C. In essence, BUILDMESSAGE is a hill-climbing heuristic that greedily selects those observations that, when integrated into the joint belief, result in the highest expected reward for the action a_C.

While BUILDMESSAGE is not optimal, its run time is only polynomial in the length of the observation history. The parameters of the heuristic make it applicable to both paradigms of communication that were discussed earlier. If communication has a fixed cost and the goal is simply to minimize the number of observations communicated, k can be set to t, the number of observations in the agent's observation history. This enables BUILDMESSAGE to select as many observations as needed to change the joint action to a_C, but no others. If there is a bandwidth limitation of k observations, ϵ should be close to 0, indicating that communication of up to k messages is allowed as long as there is even a marginal improvement in expected reward. Table 2 shows the new DEC-COMM-SELECTIVE algorithm, which utilizes the BUILDMESSAGE heuristic to

```
BuildMessage(L, ω_j, ε, k)
  a_NC ← arg max_a Evaluate(a, L)
  L' ← prune(ω_j, L)
  a_C ← arg max_a Evaluate(a, L')
  if Evaluate(a_C, L') - Evaluate(a_NC, L') ≤ ε
    return ∅
  else
    ω_C ← ∅
    while (|ω_C| ≤ k) ∧ (a_NC ≠ a_C)
      v_MAX ← −∞
      for each ω ∈ ω_j
        L' ← prune(ω, L)
        v ← Evaluate(a_C, L')
        if v > v_MAX
          v_MAX ← v
          ω_MAX ← ω
      ω_C ← ω_C ∘ ⟨ω_MAX⟩
      L ← prune(ω_MAX, L)
      ω_j ← ω_j − ω_MAX
      a_NC ← arg max_a Evaluate(a, L)
    return ω_C
```

Table 1. The BuildMessage heuristic greedily selects the observations that lead to the greatest increase in expected reward for a_C, the action that would be executed if the agent communicated its entire observation history.

choose when and what to communicate. It is invoked in any timestep when a particular agent is allowed to communicate (i.e. there is bandwidth available for it to use in this timestep).

```
Dec-Comm-Selective(L^t, ω_j^t, ε, k)
  ω_C ← BuildMessage(L^t, ω_j^t, ε, k)
  if |ω_C| > 0
    communicate ω_C to teammates
    L^t ← prune(ω_C, L^t)
    ω_j^t ← ω_j^t − ω_C
  if message ω_i^t was received from teammate i
    L^t ← prune(ω_i^t, L^t)
  a ← arg max_a Evaluate(a, L^t)
  take action a
  receive observation ω_j^{t+1}
  ω_j^{t+1} ← ω_j^t ∘ ⟨ω_j^{t+1}⟩
  L^{t+1} ← ∅
  for each L_i^t ∈ L^t
    L^{t+1} ← L^{t+1} ∪ growTree(L_i^t, a)
  return [L^{t+1}, ω_j^{t+1}]
```

Table 2. One time step of the Dec-Comm-Selective algorithm for an agent j

5 Experimental Results

5.1 Multi-agent Tiger Domain

The multi-agent tiger problem [2] is a two-agent extension to the classical tiger problem [17]. This domain is comprised of a room with two doors. Behind one door is a tiger, and behind the other is a treasure. Each agent may either choose to open a door or to perform LISTEN, an information-gathering action that provides a noisy observation about the position of the tiger. The goal of the problem is to avoid the tiger and instead to open the door hiding the treasure.

To make this an interesting benchmark for multi-agent systems, an explicit coordination problem is built into the domain. The maximum reward is obtained when both agents simultaneously open the door with the treasure. A penalty is incurred when both agents open the door with the tiger. However, the worst penalty occurs when each agent opens a different door. This coordination problem requires the agents to consider the actions of their teammates when making their own decisions.

Our experimental results demonstrate that the DEC-COMM-SELECTIVE algorithm enables a team of agents to make execution-time communication decisions not only about **when** to communicate, but also about **what** to communicate, ensuring that they do not send unnecessary information. Table 3 summarizes the results of the experiment. We generated centralized a policy for the team using the Cassandra POMDP solver [15]. We then ran 1000 trials each of the DEC-COMM and DEC-COMM-SELECTIVE algorithms, allowing the team to execute for 6 timesteps in each trial. The DEC-COMM-SELECTIVE algorithm enables agents to broadcast almost 30% less observations with only a small reduction in performance.

	Average Reward	Average # Communications
FREE COMMUNICATION	11.95	10.0
DEC-COMM	9.35	5.14
DEC-COMM-SELECTIVE	8.41	3.68

Table 3. Results for the tiger problem.

5.2 Colorado/Wyoming Domain

While the tiger domain is useful for evaluating communication strategies, in that it encodes a non-transition independent coordination problem in which agents must act jointly to maximize expected reward, it is missing other characteristics that are necessary to illustrate the full range of communication decisions. In particular, the tiger domain has only two possible individual observations. In this paper, we introduce the Colorado/Wyoming domain that, in addition to sharing the useful characteristics of the tiger domain, also has many possible observations, and those observations have different utilities with respect to team performance.

In this domain, two agents start one of two possible 5x5 grid worlds, Colorado or Wyoming, and must meet in a predetermined location. If they are in Colorado, their goal is to meet up in Denver, at grid position (2,4). If the agents are in Wyoming, they must rendezvous in Cheyenne, located at grid position (5,5) (see Figure 1). Each agent can move NORTH, SOUTH, EAST, or WEST, with each move succeeding with probability $p = 0.95$ and incurring a cost of -1. An agent can also STOP or send up a SIGNAL. Similarly to the multi-agent tiger domain, the Colorado domain contains an explicit coordination problem. If both agents are at the correct goal location when they simultaneously send up a SIGNAL, they receive a joint reward of +20. If they send up simultaneous SIGNALs from an incorrect location, they receive a reward of -50. However, if only one agent SIGNALs, or if they signal in different locations, the team incurs a penalty of -100.

Fig. 1. Figure (a) is one possible configuration of the two agents in Colorado, with the goal, Denver, at (2,4). Figure (b) is one possible configuration of the two agents in Wyoming, with Cheyenne located at (5,5).

In order to progress toward the correct goal location, the agents must observe their environment. Both Colorado and Wyoming contain flat and mountainous regions. However, the probability that an agent will observe MOUNTAIN in Colorado is slightly higher than observing it in Wyoming. Likewise, the observation PLAIN is more probable in Wyoming. Colorado and Wyoming also contain distinctive tourist attractions. It is somewhat likely that an agent will see a sign for PIKESPEAK in Colorado or a sign for OLDFAITHFUL in Wyoming, but very unlikely that these would be observed in the opposite state. Because an agent is much more likely to observe PIKESPEAK in Colorado than in Wyoming, but only slightly more likely to see a MOUNTAIN, it is clear that a PIKESPEAK observation would be more valuable to communicate to a teammate.

In our experiment, we demonstrate that the BUILDMESSAGE heuristic is able to identify and choose to communicate important observations. We compared its performance to the performance achieved by choosing random observations. A centralized policy for the team was generated using the Q-MDP heuristic [16]. We ran 1000 trials of each heuristic, with 10 timesteps per trial. The bandwidth limitation that we applied allowed agents to communicate one observation every two timesteps. Table 4 shows the results of the experiment. The BUILDMESSAGE heuristic clearly outperforms a random selection of observations, demonstrating that it successfully identifies observations that have high value of information.

	Average Reward	Average # Communications
HEURISTIC	4.70	4.29
RANDOM	0.94	4.56

Table 4. Results for the Colorado/Wyoming problem.

We alse performed an experiment to demonstrate the utility of our approach even in domains in which agents can operate independently. In this experiment, we added an absorbing state to the domain. Each agent transitions to that state when it SIGNALs. Reward is additive, with no requirement that agents SIGNAL simultaneously. This is a problem that can be solved with independent single-agent POMDPs. However, as the results in Table 5 show, the team still benefits from communication. When agents communicate their observations to each other, they are able to solve the problem more efficiently, accruing greater reward. Our algorithm enables the team to communicate those observations that will improve team performance.

	Average Reward
INDEPENDENT POMDPS	3.78
DEC-COMM-SELECTIVE	4.23
FREE COMMUNICATION	5.27

Table 5. Mean discounted reward for the modified Colorado/Wyoming problem.

6 Conclusions and Future Work

This paper discusses the need to reason about **what** to communicate when coordinating a multi-agent team. We identify two paradigms of communication, and show that it is insufficient, particularly in the case where the communication paradigm is limited bandwidth availability, to reason only about **when** to communicate. We provide a polynomial-time heuristic for selecting those observations that are, within the parameters of limited communication, most valuable for team performance and demonstrate the success of this algorithm experimentally.

In this work, we make decisions about which observations to communicate. There are domains in which a finer granularity would be beneficial, where the question to be answered is which features of the state are most relevant to team performance. Factored representations operate over these state and observation features, and we intend to investigate their applicability to our work. We also intend to apply our approach to domains in which observation probabilities vary more from state to state. We believe that these domains pose an interesting challenge to the problem of reasoning about value of information.

References

1. Bernstein D S, Zilberstein S, Immerman N (2000) The complexity of decentralized control of Markov decision processes. In: Uncertainty in Artificial Intelligence
2. Nair R, Pynadath D, Yokoo M, Tambe M, Marsella S (2003) Taming decentralized POMDPs: Towards efficient policy computation for multiagent settings. In: International Joint Conference on Artificial Intelligence
3. Peshkin L, Kim K-E, Meuleau N, Kaelbling L P (2000) Learning to cooperate via policy search. In: Uncertainty in Artificial Intelligence
4. Pynadath D V and Tambe M (2002) The communicative multiagent team decision problem: Analyzing teamwork theories and models. In: Journal of AI Research
5. Roth M, Simmons R, Veloso M (2005) Reasoning about joint beliefs for execution-time communication decisions. In: International Joint Conference on Autonomous Agents and Multi Agent Systems
6. Xuan P, Lesser V, Zilberstein S (2000) Formal modeling of communication decisions in cooperative multiagent systems. In: Workshop on Game-Theoretic and Decision-Theoretic Agents
7. Goldman C V and Zilberstein S (2003) Optimizing information exchange in cooperative multi-agent systems. In: International Joint Conference on Autonomous Agents and Multi Agent Systems
8. Nair R, Roth M, Yokoo M, Tambe M (2004) Communication for improving policy computation in distributed POMDPs. In: International Joint Conference on Autonomous Agents and Multi Agent Systems
9. Emery-Montemerlo R, Gordon G, Schneider J, Thrun S (2004) Approximate solutions for partially observable stochastic games with common payoffs. In: International Joint Conference on Autonomous Agents and Multi Agent Systems
10. Doshi P and Gmytrasiewicz P J (2005) Approximating state estimation in multiagent settings using particle filters. In: International Joint Conference on Autonomous Agents and Multi Agent Systems
11. Roth M, Vail D, Veloso M (2003) A real-time world model for multi-robot teams with high-latency communication. In: International Joint Conference on Intelligent Robots and Systems
12. Bhasin K, Hayden J, Agre J R, Clare L P, Yan T Y (2001) Advanced communication and networking technologies for Mars exploration. In: International Communications Satellite Systems Conference and Exhibit
13. Rosencrantz M, Gordon G, Thrun S (2003) Decentralized sensor fusion with distributed particle filters. In: Uncertainty in Artificial Intelligence
14. Papadimitriou C H and Tsitsiklis J N (1987) The complexity of Markov decision processes. In: Mathematics of Operations Research
15. Cassandra, A R (2005) Tony's POMDP page.
At: http://www.cassandra.org/pomdp/code/index.shtml
16. Littman M L, Cassandra A R, Kaelbling L P (1995) Learning policies for partially observable environments: Scaling up. In: International Conference on Machine Learning
17. Kaelbling L P, Littman M L, Cassandra A R (1998) Planning and acting in partially observable domains In: Artificial Intelligence

A Distributed Multi-robot Cooperation Framework for Real Time Task Achievement

Sanem Sariel[1]* and Tucker Balch[2]

[1] Istanbul Technical University, Department of Computer Engineering, Istanbul, 34496, TURKEY sariel@cs.itu.edu.tr
[2] Georgia Institute of Technology, College of Computing Department, Atlanta, GA, 30332, USA tucker.balch@cc.gatech.edu

Summary. In this paper, we propose a general framework, DEMiR-CF, for a multi-robot team to achieve a complex mission including inter-related tasks that require diverse capabilities and/or simultaneous executions. Our framework integrates a distributed task allocation scheme, cooperation mechanisms and precaution routines for multi-robot team execution. Its performance has been demonstrated in Naval Mine Countermeasures, Multi-robot Multi-Target Exploration and Object Construction domains. The framework not only ensures near-optimal solutions for task achievement but also efficiently responds to real time contingencies.

1 Introduction

In this paper, we present a generic framework, **D**istributed and **E**fficient **M**ulti **R**obot - **C**ooperation **F**ramework (DEMiR-CF), designed for efficient mission achievement of a multi-robot team for inter-related tasks that require diverse capabilities and simultaneous executions. Since real world applications present additional challenges than software platforms, robustness is a key issue of a multi-robot cooperation/coordination framework. DEMiR-CF with its integrated structure can respond to several real time contingencies efficiently while dynamically maintaining high solution quality in a fully distributed fashion. In this paper, we present the generic architecture of our framework even suitable for complex mission execution in environments with failure potentialities and limited communication features (such as bandwidth limitations, limited ranges, or unexpected delays). We report the experimental results and several scenarios in the context of Naval Mine Countermeasures mission for multi-AUV coordination in [12]; an extended technical report for the framework and evaluations is provided in [13].

* Sanem Sariel is also affiliated with Georgia Institute of Technology

2 Background and Related Work

Multi-robot coordination has been an active and attractive field during the last decade because of the demand for multiple robot, UAV, UGV or rover missions, especially in military and space applications. Among different approaches, the centralized approach is not robust especially when communication is limited between operator and individual robots, and failures are highly probable. Therefore our focus is on distributed coordination frameworks in this work, and we will review literature on this subject. Parker presents one of the earlier works for distributed multi-robot task allocation, ALLIANCE, with a behavior based framework [9] for instantaneous task assignment. M+ [1] is a distributed task allocation and achievement scheme for multi-robot cooperation addressing many real time issues including plan merging paradigms. MURDOCH [4] is a framework achieving publisher/subscriber type allocation for instantaneous assignment. Dias et al. proposes a combinatorial auction based task allocation scheme: TraderBots [2]. Lemarie et al. proposes a task allocation scheme for multi-UAV cooperation with balanced workloads of robots [8]. According to [3], existing market mechanisms are not fully capable of re-planning task distributions, re-decomposing tasks, re-scheduling commitments, and re-planning coordination during execution. We would like to fill these gaps by our integrated cooperation framework. Our primary contribution in this work is the presentation of an integrated cooperation framework for a multi-robot team and the extensive design of precaution routines and solution quality maintenance schemes for single-item auctions in real time task execution. DEMiR-CF can address different types of domains, and can generate near optimal solutions even for NP-Hard problems [11] with efficient bid evaluation methods. From our point of view, task allocation, execution and contingency handling should be integrated into the cooperation framework without assuming they are achieved separately, if globally optimal solutions are desired. This is the main rationale behind our framework.

3 Distributed and Efficient Multi Robot - Cooperation Framework

DEMiR-CF is designed for complex missions including inter-related tasks that require diverse (heterogeneous) capabilities and simultaneous execution. The framework combines *distributed task allocation and coalition formation schemes* and *dynamic task selection scheme* as cooperation components and *Plan B precaution routines* some of which are implemented by *coalition maintenance/dynamic task switching scheme.* These components are integrated into one framework to provide an overall system that finds near-optimal solutions for real time task execution.

The overall objective of the robot team ($r_j \in R$, $0 < j \leq ||R||$) equipped with our framework is to achieve a mission (M) consisting of interrelated tasks

T_i $(0 < i \leq ||M||)$, by incremental assignment of all $T_i \in M$ to $r_j \in R$ while optimizing the specified objective function. Coalitions (C_i) [6] are formed to meet requirements of simultaneous executions of tasks (T_i) synchronously by a group of robots. Sizes of coalitions vary according to the required minimum number of robots ($reqno_i$) to execute the tasks.

Definition 1. *(executable task): T_i is an executable task, if at least $reqno_i$ number of robots can be assigned for execution.*

An example of such a task may be pushing a heavy object requiring more than one robot. Tasks are preemptive: the activity of task execution can be split during runtime if another advantageous situation arises or environmental conditions impel.

Definition 2. *(candidate task and suitable robot) T_i is a candidate task for the robot r_j if the $reqcap_i$ is a subset of the cap_j and the precedence constraints of the task are satisfied; r_j is a suitable robot for the task T_i.*

Coalitions are formed by *suitable robots*. For the robots to be prepared for the contingencies, models of the system tasks and other robots are kept in each robot's world knowledge as corresponding FSMs. Task states are: *free*, *auctioned*, *being executed*, *achieved*, *uncertain* (interpreted as state *free*) and *invalid*. Robot states are: *idle*, *executing*, *failed* and *auctioneer*. The state transitions of FSMs are activated by either own motivations or incoming information from other robots. *Model Update Module* is responsible for checking and updating robot's own models. All modules in the framework and information flow among them are given in Figure 1. *Model Update*, *(System) Consistency Checking and Dynamic Task Selector* modules perform *Plan B precaution routines*. *Allocation scheme* ensures distributed task allocation. *Coalition scheme* implements synchronized task execution and *coalition maintenance* procedures. A sample flow of the operations in the framework is summarized as:

1. Mission task definitions are given to the robots (time-extended representation of tasks with precedence constrains to achieve overall mission).
2. Each robot selects the most suitable candidate task to execute by global cost consideration among mission tasks (dynamic task selection/switching).
3. Corresponding robots offer auctions for the selected tasks. In auctions, inconsistencies and conflicts are resolved.
4. Coalitions are formed for the announced tasks making sure that each robot is in the most suitable coalition from global solution quality point of view.
5. During task execution, simultaneously, dynamic task selecting/switching mechanism ensures to switch between tasks, if it is profitable; real time contingencies are handled. Then corresponding auction and coalition formation procedures (2-4) are applied continually.

Real time situations in which task switching is necessary are given in the next section.

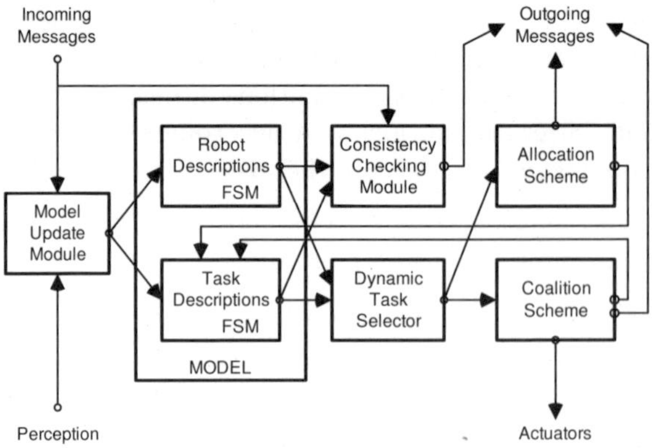

Fig. 1. DEMiR-CF Modules

3.1 Real Time Issues and Requirements

Since the world is beyond the control of the robots and change continuously in real world applications, the difficulty of multi-robot task execution problem goes beyond the task allocation problem. In particular, multi-robot systems deal with difficulties arising from noisy sensor information, unexpected outcomes of actions, environmental limitations (especially in communication) and presence of failures of hardware. All these factors may affect the overall solution. We list evolving circumstances that may change the solution as:

1. Own failure detection: Robots detect their own failure.
2. Failure detection of another robot: Robots detect another robot's failure.
3. Change in the estimated task execution cost/time: Environmental dynamics, uncertain knowledge, or hardware problems may cause delays on task execution or early achievements of tasks. Uncertain sensor and/or localization information may also result in incorrect estimations.
4. Change in the task definitions: Task dependencies, priorities, or the objective (goal) may change. Some tasks may become invalid during runtime.
5. New online tasks may be given by human operators or discovered by robots themselves.
6. New robots may be released, or some failed robots may be repaired or may recover from trap like threats.
7. Intervention and manual changes on assignments by external agents.

Some of these situations may arise after either internal or external events. Given these contingencies, even solution of an approach capable of finding optimal solutions may become sub-optimal under uncertainties of real world applications. Verification of the solution optimality is a difficult issue for real world applications. Therefore in the last decade researchers proposed effective approaches, opportunistic methods without giving boundaries on the overall solution quality except [7]. However, their work assumes perfect communication and contingencies are not considered in the boundaries. For now, these boundaries are given for the complete information cases.

DEMiR-CF is designed as being capable of dealing with the situations presented above. The framework can efficiently respond to these events and solution quality is maintained simultaneously with real time task execution.

3.2 Task Representation

Tasks are represented by a data structure containing information regarding the task execution requirements and the task status. Tasks are represented as septuples $< id, type, reqcap, deplist, reqno, relinfo, precinfo >$. System generated task ids are generated initially before mission execution and common for all robots. However online task ids may be different for each robot. Robots have initial knowledge about task types (*type*) and corresponding execution methods before mission execution. Requirements (*reqcap*) define special sensors and capabilities required to execute the task. Dependencies (*deplist*) are represented with hard and soft dependent task ids. We define two types of dependencies for representing precedence relations. Hard dependency implies sequential execution while soft dependency allows parallel execution [10]. Minimum number of robots to execute the task (*reqno*) is determined. Related information ($relinfo$) represents information regarding the type such as latest location, target location, etc. Precaution information ($precinfo$) is used for contingency handling: task state, estimated task achievement time and current execution cost.

The mission is defined as an acyclic graph (not necessarily a connected graph) of inter-related tasks connected by arcs representing dependencies. An example graph representation for the object construction mission can be found in [10]. Task definitions can be changed during execution. In particular, $relinfo$, $precinfo$ and *reqno* are subject to change during execution.

3.3 Distributed Task Allocation Scheme

Task allocation and initial assignments may be carried out by using operations research methods. However, our research addresses issues of real time execution when managing the overall team by a central authority is not possible due to several real world limitations. Auction based task allocation approach is suitable to provide a scalable and efficient way of distributing tasks. Contract Net Protocol (CNP) [14] is used to select task executers. Although CNP presents the formalism on relationships between managers and contractors, it does not present details for the following questions: When should task announcements be made? How should bid values be evaluated to get globally optimal solutions? Which subset (or all) of the already allocated tasks should be re-auctioned to maintain solution quality? When should reallocations be implemented and who decides on them? Most auction based task allocation schemes offer solutions for allocating one/subset of tasks of the overall mission. However there is usually little information about when task announcements and reassignments are made. In our framework, any robot becomes an auctioneer when it intends to execute a task. Each robot selects best suitable candidate task among mission tasks by the *dynamic task selection scheme.* Basically, auction announcements are ways to illustrate intentions to execute tasks for which $reqno = 1$ or to select members of coalitions to execute tasks for which $reqno > 1$. Therefore, if more than one robot intends to execute the same task, more suitable one(s) is selected in the auction by considering cost values. Single items are auctioned and allocated in auctions. Auction negotiation implemented in the framework consists of standard steps to clear an auction. Robots can get the necessary task details from the auction offers, and then check the validity of the auction. If the auction is invalid, related precaution routines are activated. Otherwise, the candidate robot sends its cost value as a bid. The other candidate robots behave simultaneously as well. If the auctioneer cannot get the required number (*reqno*) of bids (also counting in own bid) from the other robots until the predefined deadline, it cancels the auction. Otherwise it ranks all bids and assigns the best suitable robot with the lowest cost value to the executable task (if $reqno = 1$), or suitable coalition members (if $reqno > 1$). The framework allows multiple auctions and winners for different tasks at the same time.

3.4 Dynamic Task Selection Scheme and Online Scheduling

Dynamic task selection is implemented by forming a priority queue of unachieved candidate tasks and selecting the task with the lowest cost. Priority queue is formed either by costs for executing these tasks or by considering rough schedules depending on the selected domain. If costs are the same, the priorities are considered in the given order: Robot's current task (if any), tasks already being executed, tasks awarded in auctions, and free tasks. Dynamic task switching mechanism is used by robots to switch between tasks if updates in the world knowledge compel. Therefore issues related to both online scheduling and scheduling under uncertainty are addressed.

3.4.1 Coalition Maintenance/Dynamic Task Switching Scheme

In the framework, instead of using complicated re-allocation procedures, we propose incremental selection and task switching schemes for behaving myopically while thinking globally using bid evaluation heuristics. Provided with an efficient bid evaluation heuristic, dynamic task selection scheme ensures task switching whenever it is profitable. Each robot, independent from executing a task or not, can offer another auction or select to execute a task already being executed by another robot with a worse cost value than that it will cost for itself. If task switching occurs with a coalition member, the corresponding coalition member is released from the coalition becoming a suitable robot for other tasks. When robots participate in coalitions, they are only allowed to select other tasks, when they are released from these coalitions controlled by a robot in the coalition.

3.5 Bid/Cost Evaluation

The impact of bid (cost) evaluation on the solution quality is inevitable for auction based systems, and research in this area desires more investigation. According to the taxonomy given in [5], multi-robot task allocation problems are divided into two classes based on the mission description: instantaneous vs. time-extended. Most multi-robot architectures offer solutions for instantaneous assignments. DEMiR-CF can address both types of classes by implementing incremental allocation of tasks with efficient bidding strategies. Therefore global solution quality is maintained for the mission tasks from time-extended view of the problem by means of the bid considerations. However, the approach is capable itself to offer solutions for instantaneous changes on the task description. Therefore we classify our framework capable of addressing both types of problem classes. Unless efficient bid evaluation strategies are designed, it is not possible to observe globally optimal solutions for NP-Hard problems, and additional adjustments are required to change allocations with an additional cost of communication as in combinatorial auctions. In our earlier work, we have shown that by efficient bid evaluation approach, globally near optimal solutions can be observed for auction based approach. In that work, we analyze performance of different heuristic cost functions combined with our framework for multi-robot multi-target exploration domain [11]. Incremental assignments eliminate redundant considerations for environments in which the best solution is highly probable to change, and efficient bidding strategies ensure solutions to be close to optimal with a time-extended view of the problem. Although we have shown that our approach can find near-optimal solutions for multi-robot multi-target exploration problem, we still need further investigation on bidding strategies for different domains.

3.6 Models for Contingency Handling and Plan B Precautions

In DEMiR-CF, information is not assumed to be complete. Therefore Plan B Precaution routines are embedded in the framework to enable the system to dynamically respond to various failure modes and recover from them. These precautions are taken by each robot in a distributed fashion. Current implementation use explicit communication to detect conflicts and contingencies. However failures in communication can also be handled by precaution routines. (If robots can observe each other implicitly, model updates can be implemented in a similar manner.) Related to the contingent situations, appropriate precaution routines are activated to either correct the models, or initiate a recovery. Recovery operations may include warning other robots about the problem or changing the model accordingly. These inconsistencies usually arise when robots are not informed about tasks that are achieved, under execution or under auction. To keep system consistency, robots broadcast:

- known achieved tasks in predefined time periods to prevent redundant executions. (This feature provides a bucket-brigade type of information sharing handling communication range limitations.)
- new discovered online tasks which are unachieved yet.
- task execution messages containing the updated cost value and estimated task achievement deadline information in predefined time periods as clues for the executer robot is still alive and the task is under execution.
- task achievement message when the task is achieved.
- cancellation message if task execution is cancelled.
- task invalidation message when invalidities are detected.

Precaution routines are given in Table 1. Most of the contingencies are detected by checking models, and model updates are implemented (Table 2-3). One standard way of detection of robot failures is sending heart-beat signals. However in our framework, incoming messages from other robots are taken as clues for running properly. More complicated prediction models may be used for more accurate failure prediction. Some misleading beliefs such as setting state of a robot as *failed* although it is running properly may cause parallel executions. This is a desired feature for the mission completion point of view. Designed precautions resolve these kinds of inconsistencies if communication resources permit in later steps. In the design of precautions, it is assumed that robots are trusted and benevolent.

Table 1. Precautions for Contingencies and Conflicts

Contingency or Con ict by inconsistencies	Precaution
Any message from an unrecognized system robot is received.	Robot model is created with the corresponding state derived from the message.
Any message related to an unrecognized task is received.	Task is added to the task list with the corresponding state.
An already achieved task is announced as a new task/being executed/cancelled/auctioned.	Warning message is sent to the sender.
A task being executed/auctioned is announced as being executed/auctioned.	Only the robot with the minimum cost continues to the operation.
Cancellation message is received for a task already being executed by own.	Robot state is set "idle".
A cancellation is message is received for a task being executed by the sender robot.	Task and robot states are set as "free" and "idle", respectively.

Table 2. Model Checking for Tasks and System Robots

Status	**Action**
The time duration from the latest communication with a robot is longer than the threshold.	Robot state is set as "failed". Related task state is set as "uncertain".
Task in execution is not achieved although the estimated deadline is reached.	Task state is set as "uncertain".
Task state is "auctioned" for longer than predefined time period.	The task state is set as "uncertain".

Table 3. Model Updates Related to The Messages

Message Type	**Action**
Any type	Current time is registered as the latest comm. time with the robot and for the task.
"achieved" - valid	The robot and task states are set as "idle" and "achieved", respectively. If the task is in consideration (in schedule or in execution), it is cancelled.
"execution" - valid	If there are other tasks with state "being executed by this robot", states are changed as "uncertain".

4 Evaluation of DEMiR-CF

Our framework is evaluated in three different domains: Object construction [10], Multi-robot multi-target exploration [11] and Multi-AUV Naval Mine Countermeasures domains [12]. First two evaluations are implemented on an abstract simulator, while the third one is on the realistic US NAVY simulator. Readers are referred to the corresponding papers and to the extended technical report [13] for the performance evaluations of the framework. From general point of view, there is a tradeoff between maintaining high solution quality and the increasing communication and continuous bid evaluation requirements. According to the metric defined in [5], computational requirements per task is $O(1)/bidder$ and $O(m)/auctioneer$, when the number of robots is m. Before selecting the best bidder, each robot selects the best suitable task for itself and offers an auction for the task from time-extended point of view of the mission. Therefore among n tasks, assuming bid evaluation complexity is $O(l)$ for each task, the selection is implemented in $O(nl)$. If $nl \gg m$, this bound is given as $O(nl)/task$. To ensure system solution quality, robots continuously evaluate bids for the unachieved tasks and dynamically switch among tasks, if it is profitable, one of the differences of our framework from others, in each time step (the worst case). For standard task allocation, communication complexity is $O(nm)$ under normal circumstances. To maintain high solution quality, precaution messages are sent. However, still the complexity is given with the same bound (multiplied by an additional scalar). The performance of DEMiR-CF is shown to be bounded by 2*OPT for the Multi-robot multi-target exploration domain [11].

5 Conclusions

In this work, we present our generic cooperation framework, DEMiR-CF, for multi-robot teams. The framework combines a distributed auction based allocation method and several precaution routines to handle contingencies and communication limitations of real world domains and to maintain high solution quality with available resources. DEMiR-CF is evaluated for different domains. Near future work include further evaluations of the framework for different complex domains and specifying design issues for these domains.

References

1. Botelho SC and Alami R (1999) M+: a scheme for multi-robot cooperation through negotiated task allocation and achievement. IEEE Intl. Conf. on Robotics and Automation
2. Dias MB, and Stentz A (2002) Opportunistic Optimization for Market-Based Multirobot Control. IEEE/RSJ Intl. Conf. on Intelligent Robots and Systems
3. Dias MB, Zlot RM, Kalra N, and Stentz A (2005) Market-Based Multirobot Coordination: A Survey and Analysis, Robotics Institute. Carnegie Mellon University, Tech Report, CMU-RI-TR-05-13
4. Gerkey G and Matric MJ (2002) Sold!: Auction Methods for Multirobot Coordination. IEEE Trans. on Robotics and Automation, vol. 18 no.5, pp. 758-768
5. Gerkey B and Mataric MJ (2004) A Formal Analysis and Taxonomy of Task Allocation. Intl. Journal of Robotic Research, 23(9): 939-954
6. Horling B and Lesser V (2005) A Survey of Multi-Agent Organizational Paradigms. The Knowledge Engineering Review, 19(4):281-316
7. Lagoudakis MG, Berhault M, Koenig S, Keskinocak P and Kleywegt AJ (2004) Simple Auctions with Performance Guarantees for Multi-Robot Task Allocation. IEEE/RSJ Intl. Conf. on Intelligent Robots and Systems
8. Lemarie T, Alami R, and Lacroix S (2004) A Distributed Task Allocation Scheme in Multi-UAV Context. IEEE Intl. Conf. on Robotics and Automation
9. Parker LE (1998) ALLIANCE: An Architecture for Fault Tolerant Multi-Robot Cooperation. IEEE Trans. on Robotics and Automation, 14(2): 220-240
10. Sariel S and Balch T (2005) Robust Multi-Robot Coordination in Noisy and Dangerous Environments. Georgia Institute of Technology, College of Computing, GVU, Tech Report GIT-GVU-05-17
11. Sariel S and Balch T (2006) Efficient Bids on Task Allocation for Multi-Robot Exploration Problem. The Nineteenth International FLAIRS Conference
12. Sariel S, Balch T and Stack JR (2006) Empirical Evaluation of Auction-Based Coordination of AUVs in a Realistic Simulated Mine Countermeasure Task. The 8th Intl. Symposium on Distributed Autonomous Robotic Systems (DARS)
13. Sariel S, Balch T and Stack JR (2006) Distributed Multi-AUV Coordination in Naval Mine Countermeasure Missions. Georgia Institute of Technology, College of Computing, GVU, Tech Report GIT-GVU-06-04
14. Smith RG (1980) The Contract Net Protocol: High level communication and control in a distributed problem solver. IEEE Trans. on Computers, C-29(12): 1104-1113

Empirical Evaluation of Auction-Based Coordination of AUVs in a Realistic Simulated Mine Countermeasure Task

Sanem Sariel[1]*, Tucker Balch[2], and Jason Stack[3]

[1] Istanbul Technical University, Department of Computer Engineering, Istanbul, 34496, TURKEY sariel@cs.itu.edu.tr
[2] Georgia Institute of Technology, College of Computing Department, Atlanta, GA, 30332, USA tucker.balch@cc.gatech.edu
[3] Naval Surface Warfare Center, Panama City, FL, 32407 USA Jason.stack@navy.mil

Summary. In this work, we evaluate performance of our distributed cooperation framework, DEMiR-CF, for Naval Mine Countermeasure missions on the US NAVY's ALWSE-MC simulator against different contingencies that may arise run time. Our cooperation framework integrates a distributed task allocation scheme, coordination mechanisms and precaution routines for multi-robot team execution. Its performance has been demonstrated in Multi-robot Multi-target exploration and Object Construction domains. Marine applications provide additional challenges such as noisy communication, position uncertainty and the likelihood of robot failures. There is a high probability that the initial assignments are subject to change during run time, in these kinds of environments. Our framework ensures robust execution and efficient completion of missions against several different types of failures. Preliminary results for MCM missions are promising in the sense of mission completion, and AUV paths are close to optimal in the presence of uncertainties.

1 Introduction

Undersea operations using AUVs (Autonomous Underwater Vehicle) provide a different and in some ways a more challenging problem than tasks for UAVs and UGVs. In particular, communication windows are restricted and bandwidth is limited. Coordination among agents is correspondingly more difficult. In traditional approaches, a central planner initially assigns subtasks for a set of AUVs to be performed in achieving the team goal. However, initial assignments of tasks may become inefficient during real time execution due to the

* Sanem Sariel is also affiliated with Georgia Institute of Technology

real world issues (e.g. failures), and these allocations are subject to change if efficiency is a high concern. Therefore reallocations are needed and should be performed in a distributed fashion. To facilitate this flexibility, we offer a distributed auction based cooperation framework, **D**istributed and **E**fficient **M**ult**i** **R**obot-**C**ooperation **F**ramework (DEMiR-CF) [8], an online dynamic task allocation (reallocation) system to achieve a team goal while using resources effectively with integrated task scheduling and execution capabilities, that can also respond to and recover from real time contingencies such as communication failures, delays, range limitations and robot failures. DEMiR-CF has been implemented and tested extensively in the multi-robot multi-target exploration domain [7]. In this paper, we report performance of our framework against realistic difficulties in multi-AUV coordination for Naval Mine Countermeasure (MCM) mission on the US Navy's Autonomous Littoral Warfare Systems Evaluator- Monte Carlo (ALWSE-MC) simulator [1].

2 Background and Related Work

DEMiR-CF is a distributed mechanism for real time task execution and designed to use advantages of auction based approaches and to integrate additional routines for solution quality. Other efficient works in auction based coordination research are: M+ [2], MURDOCH [5], TraderBots [3] and Lemarie's allocation scheme [6]. According to the review given in [4], existing auction based systems are not fully capable of re-planning task distributions, re-decomposing tasks, re-scheduling commitments, and re-planning coordination during execution. Our approach aims at filling these gaps. We propose an integrated cooperation framework for multi-robot task execution, and here in this paper, we analyze performance of precaution routines and solution quality maintenance schemes for single-item auctions in a multi-AUV coordination context. Experiments are performed in a realistic simulation environment with real time constraints and events such as AUV failures, communication range limitations, failures and delays. Precaution routines embedded in the framework not only recover from failures but also maintain the high solution quality. With an efficient bid evaluation approach, the framework provides near optimal solutions [7]. Our experiments show that communication delays significantly impact the solution quality and should be analyzed in multi-robot systems especially working in harsh environments. As experiments and scenarios demonstrate, online task handling performance of the framework with task switching mechanism is promising.

Naval mine countermeasures (MCM) are actions taken to counter the effectiveness of underwater mines. MCM operations include finding and seizing mine stockpiles before they are deployed, sweeping desired operational areas, identifying mined areas to be avoided, and locating and neutralizing individual mines [10]. Our research is focused on the subset of MCM that involves locating and mapping all individual mines in an operational area. In general, recognizing proud mines on the seafloor is not overly difficult; the difficulty arises with the abundance of non-mine objects on the seafloor that possess mine-like characteristics (*e.g.*, geologic outcroppings, coral, manmade debris, etc.). This ample supply of false alarms has necessitated the following strategy typically employed by the Navy: detect and classify the mine-like objects (MLOs) with high-coverage rate sensors (*e.g.*, sidelooking sonar), employ advanced signal processing techniques for maximal false alarm reduction, then revisit the remaining MLOs with identification-quality assets (*e.g.*, electro-optic sensors) to confirm them as mines or dismiss them as false alarms. It is this strategy which the research proposed herein attempts to implement in a distributed, near optimal fashion. Achieving this mission with an AUV team requires effective task allocation mechanisms and several precautions.

3 The DEMiR-CF Framework for Naval MCM Missions

DEMiR-CF is designed for complex missions including inter-related tasks (with precedence constraints) that require diverse agent capabilities and simultaneous execution. The framework combines *distributed task allocation and coalition formation schemes*, and *dynamic task selection scheme* as cooperation components, and *Plan B precaution routines* some of which are implemented by *dynamic task switching scheme.* These components are integrated into one framework to provide an overall system that finds near optimal solutions for real time task execution. The overall objective of the robot team ($r_j \in R$, $0 < j \leq ||R||$) equipped with our framework is to achieve a mission (M) consisting of interrelated tasks T_i ($0 < i \leq ||M||$), by incremental assignment of all $T_i \in M$ to $r_j \in R$ while optimizing the specified objective function. Details of DEMiR-CF are provided in [8], and an extended version of the overall framework and the implementation details given in this paper is provided as a technical report [9]. In this paper, we report experimental evaluations of our framework and details about application of the framework for a real mission execution. The reference mission in this research is to detect, classify, and identify underwater mines in a given operational area simulated in a PC-based software, ALWSE-MC [1], analysis package designed to simulate multiple autonomous vehicles performing missions in the littoral regions including mine reconnaissance, mapping, surveillance, and clearance. This mission employs two types of vehicles: unmanned underwater vehicles (UUVs) which are free swimming AUVs and possess large-footprint sensors (*e.g.*, side-scan sonar) for detection and classification (D/C) of mines and seafloor crawlers equipped with short-range, identification-quality sensors (*e.g.*, camera). The crawlers have the ability to stop at an object and take a picture with a camera.

Our general task representation is designed as being capable of representing complex tasks with inter-dependencies. In particular, in this case study, tasks do not have interdependencies. Two types of tasks are defined for vehicles: "visit waypoint" (w) and "identify MLO" (t). In the task representation, required capabilities are represented for each type of task: $reqcap_w$ contains side-scan sonar and $reqcap_t$ contains cameras besides the standard capabilities of AUVs common in both types of vehicles. The coverage mission (M_C) contains predefined number of waypoints ($w_i \in M_C$, $0 < i \leq ||M_C||$) to be visited by all UUVs ($R_{UUV} \subset R$). One way of task representation is to directly assign tasks for each waypoint. However this representation has a drawback of high communication requirements for efficient completion of the mission. Instead, we represent waypoint tasks as interest points of regions/search areas ($W_k = \cup w_i$, $\forall w_i$ is unvisited, and $W_k \subseteq M_C$). These regions (and corresponding centers) are determined by robots during runtime dynamically. The advantage of this representation is that the only information necessary to negotiate over tasks contains the interest point information, providing data compression. Regions determined by different UUVs may vary during runtime and sometimes overlap. However, the uncertainty related to the tasks is within an acceptable degree compared to the requirements of complete knowledge sharing. Before defining regions, relative distance values, $reldist(r_j, w_i)$, are determined for each unvisited waypoint w_i as in Eq. 1, where $dist$ function returns the Euclidian distance between points. r_k locations are the latest updated locations of the robots. If there is no known active robot, $reldist(r_j, w_i)$ value is taken as only the own distance.

$$reldist(r_j, w_i) = dist(r_j, w_i) - min_{\forall k \neq j}(dist(r_k, w_i)),\ r_k \text{ is active} \quad (1)$$

Each robot defines its regions (W_{jk}, $1 \leq k \leq ||R_{UUV}||$) number of which equals to the number of UUVs believed to be running properly. After sorting $reldist(r_j, w_i)$ values of unvisited waypoints, the regions are determined on the sorted list, containing approximately same number of waypoints. The first region is the region that the robot has the highest interest (but negotiations are needed to resolve conflicts if there is another UUV with a similar interest). The identification mission (M_I) contains unknown number of tasks for MLO locations ($t_i \in M_I$, $0 < i \leq ||M_I||$) to be visited by crawlers. Therefore the tasks in M_I are generated online during runtime. For bid evaluations, we use heuristic functions proved to provide close to optimal results for multi-robot multi-target domain [7]. These cost functions, explained in the next section, provide time extended consideration of tasks for instantaneous assignment with a tractable and efficient way. A conceptual flowchart summarizing operations of UUVs and crawlers, and the general operations implemented by both types of AUVs is given in Fig. 1.

3.1 Exploration for Detection and Classification of MLO Locations

To begin the mission, the UUVs survey the operational area following waypoints determined *a priori*; however, corresponding regions containing waypoints may be reassigned by negotiations among UUVs autonomously. After determining regions, each UUV offers an auction for the highest interested region for itself and offers its selected interest point information as an auction. After negotiations on several auctions, each UUV is assigned to the closest

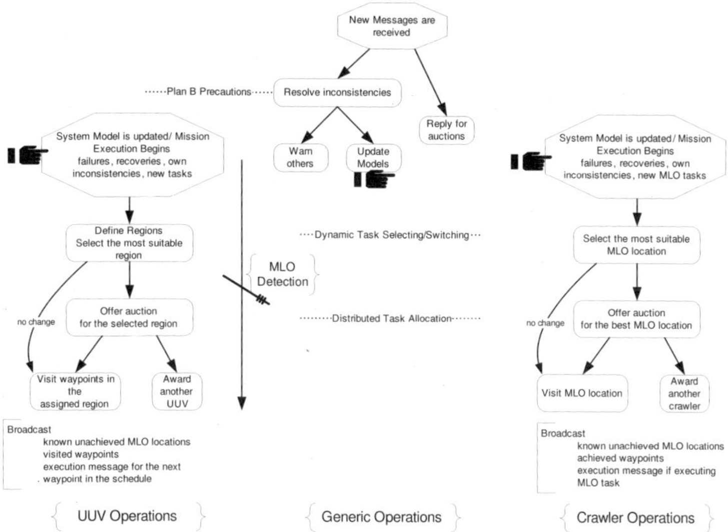

Fig. 1. Conceptual Flowchart related to the AUV Operations

region (interest point). If more than one robot is at the same distance to the interest point, the one with the minimum id is assigned. The other UUVs continue to offer auctions for the remaining regions. Allocations of the regions may also change during run time to maintain solution quality. Whenever UUVs detect UUV failures or recoveries from failures they change their region definitions accordingly and offer new auctions. After region assignments are implemented, each robot visits waypoints in the corresponding region (W_j) by ordering them descendingly according to their cost values as in Eq. 2.

$$\begin{aligned} c(r_j, w_i) &= \alpha * dist(r_j, w_i) + (1-\alpha) * [dist(w_{f1}, w_{f2}) \\ &\quad - max(dist(w_i, w_{f1}), dist(w_i, w_{f2}))] \\ \{dist(w_{f1}, w_{f2}) &= max(dist(w_k, w_l)),\ w_{i,k,l,f1,f2} \in W_j\} \end{aligned} \quad (2)$$

This heuristic function considers boundary targets, w_{f1} and w_{f2} in W_j which are the targets having the maximum distance value. The basic idea of this function is that these targets determine the diameter of the region (W_j) and both of them should be visited. This heuristic method forwards robots to these farthest targets within their area to some degree. By introducing a constant (α), this degree can be adjusted and it is taken as $2/3$. This heuristic function produces close to optimal results for multi-robot multi-target domain [7]. If there are more than one pair of boundary targets, the pair of which has a member with the smallest distance to the UUV is selected.

As UUVs detect the MLOs on their way, they broadcast these estimated target positions to all AUVs (*i.e.*, tasks for crawlers are generated online).

Then MLO information can propagate (in bucket-brigade fashion) to all other AUVs in the group that can possibly be reached. Periodic broadcasting of important information (coming from either own sensors or external agents) is a way to handle communication range limitations.

3.2 Identification of MLOs

When crawlers are informed about MLO locations, they update their world knowledge and dynamically select the best MLO target to visit and offer auctions. Therefore they can switch among tasks when new tasks appear, if it is profitable. It is also possible that a crawler may inadvertently discover a mine without being informed of its position by a UUV. In this case, the crawler identifies the target, adds it to its task list as an achieved task, and broadcasts achievement information for maintaining system consistency. Crawlers determine their bid values by Eq. 3, where t_k is the closest unvisited MLO target to t_i. This cost function provides a greedy look ahead for visiting MLO targets rather than only considering the distances between target and the AUV. An additional penalty is applied to the cost, if there is another profitable alternative way of visiting tasks.

$$c(r_j, t_i) = \begin{cases} dist(r_j, t_i) + dist(t_i, t_k) - dist(r_j, t_k)\ , & \text{if } (dist(t_i, t_k) > dist(r_j, t_k)) \\ dist(r_j, t_i) & \text{otherwise} \end{cases} \quad (3)$$

In the identification task, when crawlers are within an area close to a MLO location, they begin keeping time while surveying the MLO location. Whenever the time limit is reached, they set the task status as achieved and broadcast this information. If there is detection during this time period, MLO location is considered as an actual mine and task achievement is directly applied, otherwise it is determined as a false alarm after deadline. In either case, the task is achieved.

4 Experimental Results

Performance of our framework and precaution routines is evaluated in ALWSE-MC. Three sample scenarios in the simulation are given to illustrate performance of our framework for Naval MCM missions. UUVs are equipped with sensors capable of detecting mines within 30 feet from skin of target. However they are not able to correctly identify them. Crawlers are equipped with cameras which can both detect and identify mines within 20 feet. None of the AUVs have certain search patterns. UUVs have internal navigation errors therefore their estimated location values are different from actual locations in most cases. Two AUVs can communicate each other whenever the receiver AUV is in the sender AUV's transmitter range, within its transmitter beam width, and sender AUV is within transmitter AUV's receiver beam width.

All UUVs and crawlers begin execution from a deployment area. There is no *a priori* information about mine locations. 121 waypoint locations (environment size: 200x200) are known but are not assigned initially. UUVs begin negotiations and divide the overall mission area into three (known number of UUVs) regions. Since they are within line of sight, they can communicate their location information. Therefore initially defined regions are nearly the same for all UUVs. Fig. 2 illustrates a successful mission scenario with three UUVs and two crawlers. Allocations of waypoints after negotiations can be seen in Fig. 2(b). Since there are no failures, waypoint assignments do not change during run time. However crawlers sometimes switch among tasks if they are not informed about tasks that are being executed. And sometimes parallel executions occur. Whenever they are in communication range, they can resolve the conflicts efficiently by means of the precaution routines. As in Fig. 2(a), crawlers can also detect mines without being informed. Routes of the crawlers may seem somewhat random. However it should be noted that tasks related to the MLO locations appear online during run time when they are discovered, and communication ranges are limited.

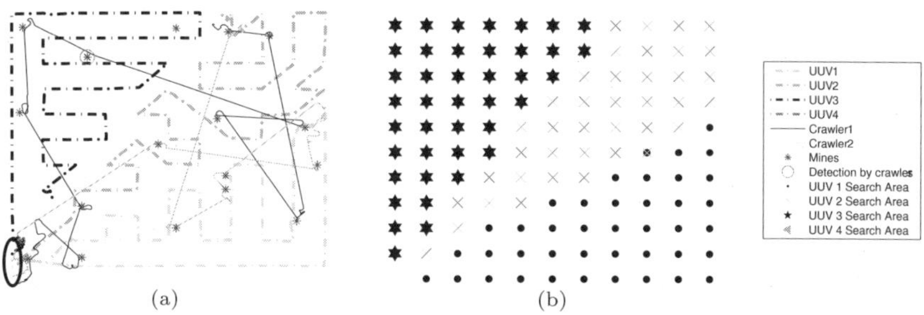

Fig. 2. Scenario 1. (a) UUVs cover the area by visiting waypoints. Crawlers visit MLO locations as they are informed. Deployment area is circled. (b) Each AUV is assigned to a region after auction based allocation of interest points.

In the second scenario, one of the UUVs fails on the same setting of scenario 1 (Fig. 3). Initial regions for all UUVs change after UUV3 fails (Fig.3 (b)). Other UUVs change region definitions and, after negotiations, they share the full area as indicated in the figure. Visited waypoints are not in their region coverage. Because of the uncertainties, some waypoints are left uncovered in schedules. However this uncertainty related problem is resolved by UUV2 and the mission is completed.

In the third scenario (Fig. 4), UUV3 fails and other UUVs detect the failure and they negotiate over the remaining unvisited waypoints and new schedules are determined as in Fig. 4(b). While these UUVs execute their tasks, another UUV (4) is released from the deployment area. Detecting a new UUV arrival, other UUVs change their region definitions accordingly (Fig. 4(d)) and offer auctions for these areas. UUV4 initially is not informed about the visited

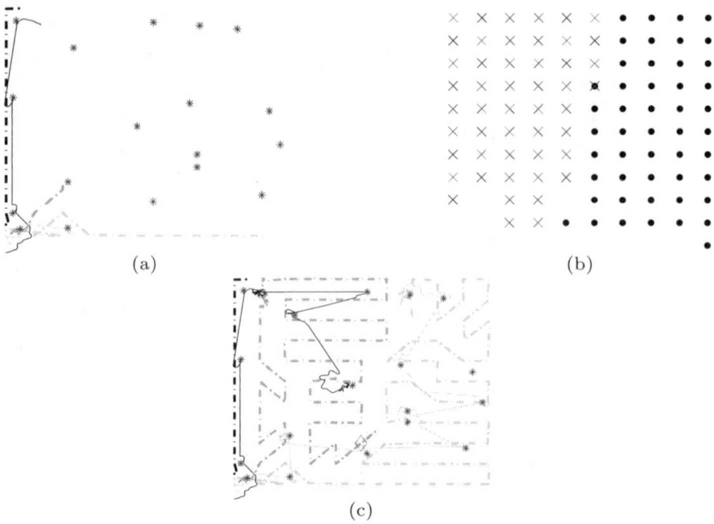

Fig. 3. Scenario 2. (a) Initially all UUVs begin execution. UUV3 fails, other UUVs take responsibility of all unvisited waypoints. (b) Region assignments are changed for UUV1-2 after detecting failure. Because of an uncertainty, one waypoint is left uncovered. (c) UUV2 completes its region coverage task, and adds the waypoint missing in (b) to its schedule after detecting that it is not visited.

waypoints and it defines its regions with this knowledge. After negotiations, the regions are assigned and schedules are formed. UUV4 redefines its regions by considering incoming information for visited waypoints.

On the same settings, experiments are conducted to evaluate message loss rate effects on mission completion success. Table 1 illustrates the results ($\mu \mid \sigma$) averaged over 10 runs. When message loss rate is different from 0, as expected, performance is degraded but linearly. It should be noted that even for rate 0.75, the overall mission (M_C and M_I) by final identification of mines is completed. Number of waypoint (w) visits increase for high message loss rates. When message loss rate is 1 there is no communication among AUVs and they cannot correctly reason about region portions. Therefore each UUV searches the full area completely. Crawlers detect and identify 12.8% of mines by their local detection in a small area (MLO target information can not be communicated in this case). Since identification is not complete, overall mission is not completed. This table illustrates performance of our framework against message losses. As a final remark, auction generation and clearing in an environment with communication delays desires special attention. Especially auction deadlines should be determined by considering communication delays which may vary during run. Plan B precautions could resolve these kinds of problems. Precautions for delayed messages on out-of-date situations prevent the system from getting into stuck into further inconsistencies and deadlocks.

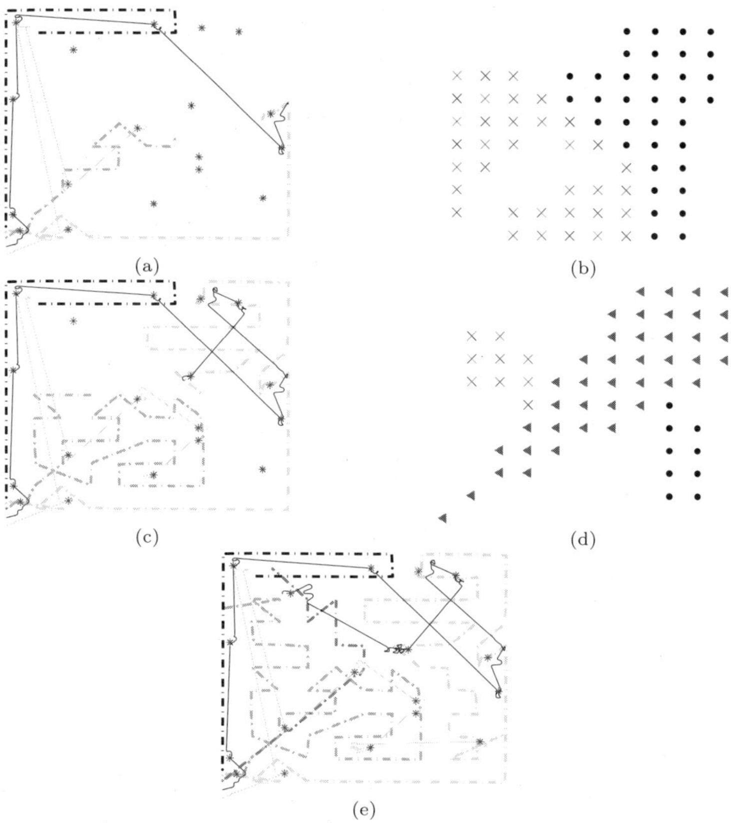

Fig. 4. Scenario 3. (a) UUV3 fails, other UUVs take responsibility of the waypoints initially assigned to UUV3. (b) Region assignments are changed for UUV1-2 after detecting failure. (c) Another UUV(4) is released from the deployment area. (d) Schedules are changed accordingly after negotiations. However UUV4 is not informed about visited waypoints and form regions by considering all waypoints. (e) After being informed about visited waypoints, UUV4 only visits unvisited waypoints in its schedule.

Table 1. Performance Results ($\mu \mid \sigma$) for Different Message Loss Rates

Mssg Loss Rate	0	0.25	0.5	0.75	1
M_C Comp. (%)	100.0 | 0.0	100.0 | 0.0	100.0 | 0.0	100.0 | 0.0	100.0 | 0.0
M_I Comp. (%)	100.0 | 0.0	100.0 | 0.0	100.0 | 0.0	100.0 | 0.0	12.8 | 4.1
M_C Comp. time	3349.4 | 60.5	3683.2 | 167.1	4909.0 | 430.1	5141.2 | 938.1	6304.2 | 139.0
M_I Comp. time	2852.8 | 35.3	3227.6 | 205.3	4205.0 | 836.9	5021.2 | 692.7	N/A
(w) first visit	1380.1 | 6.1	1390.0 | 16.3	1922.0 | 92.8	2256.6 | 334.5	2936.0 | 104.5
(w) #of visits	1.0 | 0.0	1.0 | 0.0	1.01 | 0.01	1.09 | 0.04	3.0 | 0.0

5 Conclusions

In this work, we present performance of DEMiR-CF in the context of a Naval Mine Countermeasure mission in the realistic simulator, ALWSE-MC. DEMiR-CF is a distributed framework for multi-robot teams that integrates an auction based dynamic task allocation scheme and several precaution routines to handle failures and limitations of real world task execution, and maintains high solution quality with available resources. Precaution routines can respond to several failures some of which are illustrated in the scenarios shown in this paper. Evaluations also reveal high performance of DEMiR-CF on online task and situation handling. Since the framework is a single item auction method it can be used for the environments with limited, delayed or unreliable communication. In general, the framework is designed for more complex missions of interrelated tasks. Near future work consists of more complex missions with more limitations for AUVs and task execution. It should be noted that the selected application domain, objectives and limitations are similar to the Search and Rescue (SR) domain. Therefore we believe research in this work can also be useful for different kinds of domains such as SR.

References

1. ALWSE: http://www.ncsc.navy.mil/Capabilities_and_Facilities/Capabilities/Littoral_Warfare_Modeling_and_Simulation.htm
2. Botelho SC and Alami R (1999) M+: a scheme for multi-robot cooperation through negotiated task allocation and achievement. IEEE Intl. Conf. on Robotics and Automation
3. Dias MB, Zinck MB, Zlot RM, and Stentz A (2004) Robust Multirobot Coordination in Dynamic Environments. IEEE Intl. Conf. on Robotics and Automation
4. Dias MB, Zlot RM, Kalra N, and Stentz A (2005) Market-Based Multirobot Coordination: A Survey and Analysis, Robotics Institute. Carnegie Mellon University, Tech Report, CMU-RI-TR-05-13
5. Gerkey G and Matric MJ (2002) Sold!: Auction Methods for Multirobot Coordination. IEEE Trans. on Robotics and Automation, vol. 18 no.5, pp. 758-768
6. Lemarie T, Alami R, and Lacroix S (2004) A Distributed Task Allocation Scheme in Multi-UAV Context. IEEE Intl. Conf. on Robotics and Automation
7. Sariel S and Balch T (2006) Efficient Bids on Task Allocation for Multi-Robot Exploration Problem. The Nineteenth International FLAIRS Conference
8. Sariel S and Balch T (2006) A Distributed Multi-Robot Cooperation Framework for Real Time Task Achievement. The 8th International Symposium on Distributed Autonomous Robotic Systems (DARS)
9. Sariel S, Balch T, and Stack JR (2006) Distributed Multi-AUV Coordination in Naval Mine Countermeasure Missions. Georgia Institute of Technology, College of Computing, GVU, Tech Report GIT-GVU-06-04
10. Stack JR and Manning RC (2004) Increased autonomy and Cooperation in Multi-AUV Naval Mine Countermeasures. Proceedings of Undersea Defence Technology

Principled Synthesis for Large-Scale Systems:Task Sequencing

Dylan A. Shell and Maja J Matarić

University of Southern California shell|mataric@usc.edu

Summary. This paper describes ongoing work toward a principled controller synthesis methodology for large-scale, minimalist multi-robot systems. The work's key objectives is to establish a set of programming primitives (processes) for which macroscopic behavior can be formally predicted. Such prediction is made possible by statistical physics techniques that use properties of time-invariant processes while exploiting the system's large size. This paper's focus is on the use of numerical and simulation methods during construction of the primitive process set. A computational method, developed by physicists, is used as a high-level simulation to characterize individual process behavior. The output, when interpreted qualitatively, guides distributed system design. In order to validate the approach, we consider a sequential inspection domain with a swarm of 400+ simulated robots. Synchronization is achieved through processes analyzed with the methods described, and predictions are compared with behavior exhibited in a traditional multi-robot simulation. The two simulation tools play different roles in characterizing collective behavior; the differences shed new light on the problem of multi-robot controller synthesis.

1 Introduction

Robot swarms consist of many simple, small, and inexpensive units that exploit synergistic interactions to perform tasks and achieve robustness through massive redundancy. We consider the *synthesis* problem, which requires derivation of local rules (robot control-laws and communication protocols) that enable the performance of a pre-specified global task. A method based on *composition of elementary processes* is proposed that yields controllers for large-scale systems. Most current methods are task specific, and finding a general solution to the problem remains a major challenge. This paper describes the use of simulation at different levels of detail aimed at addressing this challenge.

Researchers have begun seeking principled methods for design of minimalist systems. One approach is *analysis* of existing implementations (e.g., [5, 6]). Iterative improvements are offered for subsequent implementations and the overall-design process. Other work has focused on compilation of task-oriented rules [4], or automating the distribution of sensory information [7], to provide an *automated synthesis* methodology. Such approaches are necessarily restricted to particular tasks or automating specific capabilities, since the general problem is formidable. Analysis methods have been extrapolated to large systems (e.g.,[6, pp. 12]) but formal synthesis methods have, thus far, only considered small groups.

Our proposed synthesis method not only scales to hundreds of robots, but actually exploits large system size. Although not an automated procedure, the method makes use of predictive tools for guiding design, producing coarse descriptions of behavior for a particular class of processes. Controller design then requires processes to be combined so as to achieve task-oriented behavior at the collective level. Many theoretical questions arise when considering synthesis based on composition. Most important is the robustness of predictions of the constituent processes, because models must–of necessity–ignore some details. Together with a description of the synthesis method itself (§2) and descriptions of the simulation techniques used (§4), this paper addresses the robustness issue in experiments with a simulated swarm applied to a sequential inspection task (§3).

Our broader research deals with two key questions: 1) How feasible (and tractable) is prediction of constituent processes as a guide for system design? and 2) Given that we concentrate on a special class of computational processes, what are the computational capabilities of this class? We focus on the first question in this paper, specifically showing that the tools described are applicable to multi-robot system design for simple tasks. The second question is touched on empirically. Non-trivial capabilities are realizable with our method, as shown by considering synchronization, a canonical problem for groups of loosely-coupled asynchronous agents.

2 Analysis of individual processes

The collective behavior of a multi-robot system is difficult to predict since, in addition traditional distributed computing issues, robots have noisy sensors, imperfect actuators, and physical dynamics that constrain actions. Despite the many sources of complexity, formal synthesis and analysis methods are necessary. An important question to ask is: what information is needed *from* prediction? The level of detail required for a formal model is critically affected by the answer to this question. We accept sparse, qualitative predictions of collective behavior that ignore many low-level details. Most important from our perspective is the distinction between individual robot (*microscopic*) and group (*macroscopic*) descriptions. Explicitly modelling the system at these two levels becomes increasingly important as ever larger numbers of robots are considered.

Statistical mechanics is concerned with the derivation of bulk material properties from molecular models. The theory typically considers systems with infinitely many constituents and equilibrium methods also include other assumptions (e.g., slow changes). Simulation tools are necessary because, even with these assumptions, few models have analytical solutions. This paper is part of an ongoing research agenda to explore the limits of such assumptions, range of applicability, and appropriate generalizations for multi-robot (and multi-agent) systems. These methods require that a non-traditional view of distributed computation be considered.

A *process* describes a time-extended series of actions or events. The processes we consider result from the execution simple, local, computational rules. Each process is described by listing a range of possible states, and a (possibly non-deterministic) transition function on those states. Execution of a process is the repeated application of the transition function. In this paper, states are values of internal variables on each robot. Only homogeneous systems are considered. At some frequency the transition function maps from an existing state to a new state, based on the process state and the state of others within the communication distance. The sequence of state values generated by each robot is the *microstate evolution*; it gives the microscopic details for a process. A lower-level controller for obstacle avoidance, smoothing sensor readings, etc., has access to the robot's state variables. We are only concerned with the higher-level coordination and cooperative aspects for our processes.

We restrict the processes (and hence transition functions) to those that are *ergodic*. This implies the existence of a time-invariant probability measure on the state-space [8], since the process dynamics have a weak temporal structure. Operationally, the ergodic property allows for the time average of some quantity to be calculated by averaging over the state space.

To synthesize controllers for new problems, we envision a library of ergodic processes, each with a macroscopic description. This description is the mean of a (process-dependent) characteristic function calculated over all possible system states; it need only be calculated once for each process. Multiple processes can be combined so that, provided time-scales are chosen appropriately, the resulting behavior can be inferred from the macroscopic descriptions of the constituent processes. Function values are simply calculated over the product of the two constituent state spaces.

Ergodicity is valuable for synthesis because well-defined parts of the controller can be independently considered and each can have the ergodic property *by construction*. Using the property for system analysis requires that the overall system behavior be ergodic. Such a claim is difficult to make. (Analysis that assumes the plausibility of a "stochastic series of events" interpretation, as in [6], is far less restrictive.) Existing work that exploits ergodic dynamics does so only for part of the overall system. For example, Jones and Matarić [4] consider controllers which perform an ergodic exploration of the environment, and implicitly use this fact in predicting system performance. White et al. [9] make an assumption about the nature of the environmental dynamics that amounts to ergodicity. No work so far, however, has explicitly recognized the connection with ergodic theory.

Without inherent temporal structure, ergodic processes appear inadequate for robot controller design. Restricting design to ergodic processes certainly represents a significant shift in perspective. Typically, programming (of computers or robots) involves a decomposition of the task specification into a sequence of steps. The validation task domain explored in this paper is a sequential inspection task, and the presented controller shows that ergodicity can impose temporal ordering. The processes achieve this at a strictly macroscopic level.

3 Sequential inspection task

Given a large bounded environment with multiple sites of interest, consider the problem of having a robot swarm (in its entirety) visit these sites in a predetermined sequence. This situation arises when a mobile robot system supplements a network of static sensor nodes in order to provide higher-resolution sampling when some phenomenon is sensed or anticipated. We consider the case in which the entire swarm is tasked as a unit to perform the inspection of sites of interest in a specified order.

Collective decision making is non-trivial without a centralized leader. Synchronization of state across robots is necessary for sequential inspection because the group must collectively decide that the next site should be visited. This form of synchronization was first defined by Jennings and Kirkwood-Watts [3] in a multi-robot context. We consider individual robots that are extremely limited and show that two simple ergodic processes are sufficient for this cooperative decision making.

3.1 Process definition

The simulated robots have limited capabilities with noisy sensing and unreliable local communication. The sequential inspection controller consists of a low-level controller coupled to two ergodic processes. The processes are responsible for the decision of which site of interest to visit. The processes execute in a distributed fashion, sharing information with neighboring robots through a local-broadcast communication network (further details in §4.2).

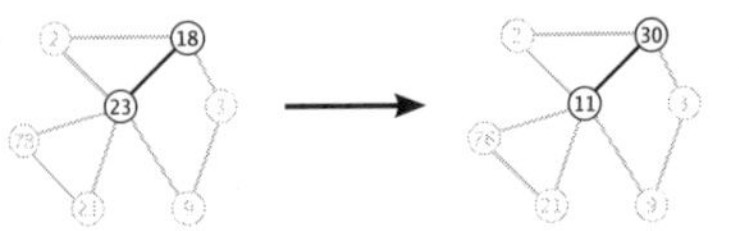

Fig. 1. Vertices represent robots and edges communication links. Numbers depict Process 1 state. A random value (12 here) is exchanged. No state information in gray is necessary, it is a strictly local interaction.

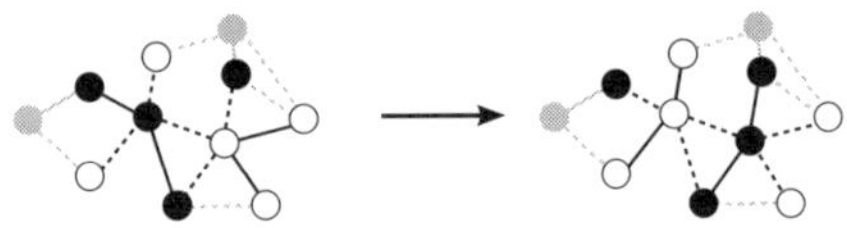

Fig. 2. A Process 2 transition. Filled vertices represent state +1, empty ones -1; solid edges depict alignment, broken edges misalignment. The number of solid (4) and broken (5) edges is conserved.

Process 1 The state of Process 1 (on each robot) is described by a positive integer. Force the following constraint to hold: the sum of the state values over all robots must equal $\mathcal{K}_1$. Then the state-space for the entire swarm consists of every possible permutation of $\mathcal{K}_1$ over the n robots. Although the summation constraint describes a global property, it is locally enforceable. Robot x in state x_i can transition to state x_{i+1} and maintain constraint by reaching an agreement with another robot, say y. Robot y must to transition from y_i to y_{i+1} so that $y_{i+1} + x_{i+1} = x_i + y_i$.

The dynamics function makes a transition from state x_i to x_{i+1} by setting apart a random portion of x_i and transmitting it to a randomly selected robot within communication range. The value of x_{i+1} is obtained by adding any portions received from neighbors. (The portion is only removed by the sender with confirmation of receipt.) Thus, the sum of state values remains $\mathcal{K}_1$. Fig. 1 shows the result of this transition.

Process 2 The second process is based on the ferromagnetic Ising model [1]. Each robot's process can be in one of two states: $\{-1, 1\}$. Two values on neighboring robots are *aligned* if they have the same state values, and *misaligned* otherwise. The total number of aligned and misaligned edges are written as N_{aligned} and $N_{\text{misaligned}}$ respectively. We write $\mathcal{K}_2 = -N_{\text{aligned}} + N_{\text{misaligned}}$.

The transition rules conserve $\mathcal{K}_2$. Two neighboring robots can calculate the effect of flipping local state (i.e., changing to the other state) by examining their immediate neighbors. This constraint can also be locally enforced. As before, the decision as to which neighbor, and whether to flip or not, is randomly selected. Process 1 is clear in terms of vertex states, Process 2 is more intuitive as operating on edges. See Fig. 2.

3.2 Process analysis

Next, consider the behavior of the processes over a range of $\mathcal{K}_1$ and $\mathcal{K}_2$ values.

Process 1 The transition function for Process 1 is a random walk of K_1 units on the robot communication graph. If robots maintain loose connectivity (i.e., disconnected subgraphs rejoin periodically) then symmetry suggest equal probabilities over each of the robots. Thus mean and variance of expected states can be calculated. Both are trivial, with analytical solutions for any values of K_1 and n. Figs. 3 and 4 compare those predictions to simulation data.

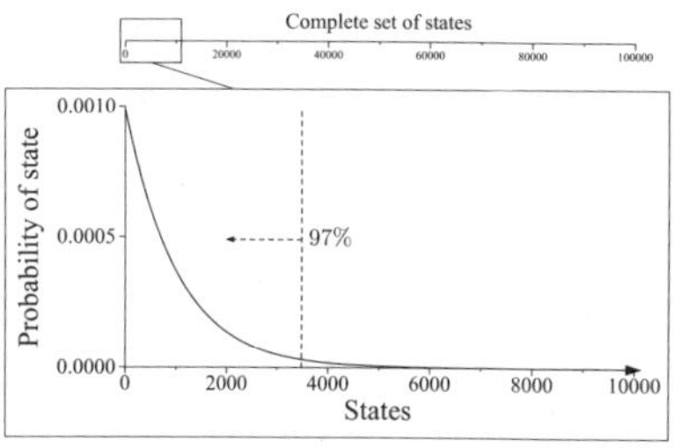

Fig. 3. The density function describing the probability of Process 1 being in a particular state at a random time. The plot of the full domain along the top and the broken line that delineates 97% of the probability mass, showing sharp peak in the distribution.

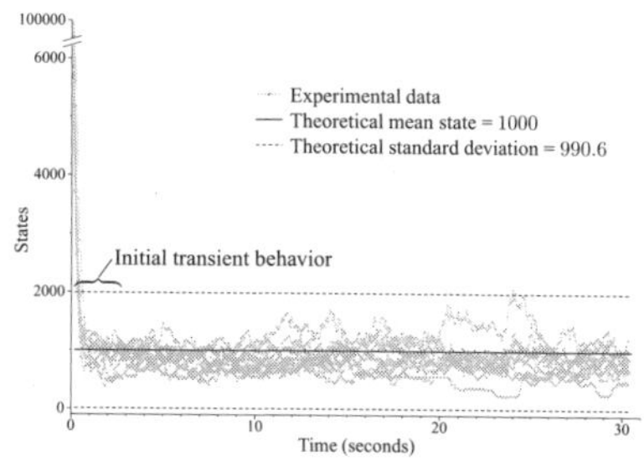

Fig. 4. Plots from ten experimental runs showing the state of a single robot ($n = 100$ and $\mathcal{K}_1 = 100000$) over time. In all cases, the plotted robot moved quickly into states well characterized by the theoretical mean and standard deviation.

Process 2 The symmetry in Process 2 is far less obvious than in the previous case. The implementation does not bias one configuration over another. Given enough time, we can expect any two configurations with the equal $\mathcal{K}_2$ to be equally likely if we assume that average network connectivity is stable (or slow changing compared with execution of Process 2 itself).

We simplify the problem by considering a model of robots placed on a 21×21 square lattice with each of the 441 robots placed on the grid and connected to nearest neighbours. To calculate the probability of an ergodic process occurring in a particular state one must construct a measure over all possible states. For n robots this means integrating over the 2^n states. Process 2's transition function was simulated using MMMC (see §4.1 for details). The mean state value M is calculated for each configuration and must lie on $[-1, 1]$. The two extremal values are given for ordered states, those midway ($M \sim 0$) are disordered. The result is a characterization of the number of given states for a given $\mathcal{K}_2$. The logarithm of this number gives S the Boltzmann entropy of the configuration (in natural units). Fig. 5 shows a construction of the entropy surface for different values of M and E. The E axis is $\mathcal{K}_2$ normalized to fall between -1 for minimum alignment and $+1$ for complete alignment. The figure shows the log conditional probability (S) of an M state given an E value.

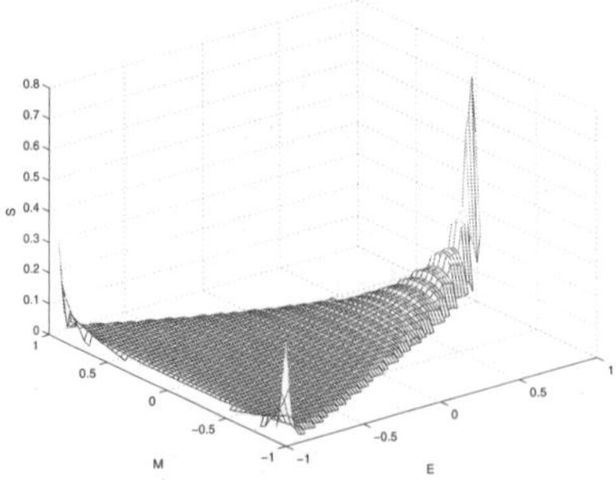

Fig. 5. The entropy S as a function of M for a given E. Since S is a logarithm of the number of states, the function is sharply peaked. This implies a high degree of certainty of the process state for extremum values of E.

For values of $E \sim +1$ there is one clear peak: the system can be expected to exhibit behavior with $M \to 0$, or an equal number of -1 and $+1$ spin values. When $E \sim -1$ the system occurs in one of two states, either many $+1$ or many -1 values. If the system is moved from a state of high E to toward low E, then it will exhibit a spontaneous symmetry-breaking phase-transition.

Coupling the two processes

Thus far, only constant $\mathcal{K}_1$ and $\mathcal{K}_2$ conditions have been considered. Now we couple the processes by allowing Process 1 to slowly increase (decrease) $\mathcal{K}_1$ provided that $\mathcal{K}_2$ decreases (increases) by the same amount. This is the same type of conservation used in defining the processes, but now applied to process parameters. Such a coupling can be modelled as a transition function operating on process parameters while maintaining $\mathcal{K}_1 + \mathcal{K}_2 = \mathcal{C}$. It is essentially a composite process with a combined state-space. We term this a *macroscopic degree-of-freedom.*

The entropy (and hence probability distribution) of the composite process can be calculated by considering the entropy function of each individual process. The exact method depends on the relative "size" of the entropy function, but a full discussion is beyond the scope of this paper. For Process 1 we can see that the first process results in a mean value of $\frac{\mathcal{K}_1}{n}$ per robot. If we permit free exchange of one unit of $\mathcal{K}_1$ for one unit of $\mathcal{K}_2$, then a decrease in $\mathcal{K}_1$ results in an increase in $\mathcal{K}_2$. With appropriate values, the decrease in $\mathcal{K}_1$ can result in a rapid transition from $M \sim 0$ to $M = -1$ or $M = +1$. This transition is the result of the average behavior of the system. We describe how this is used for the sequential inspection task below.

4 Simulation tools

Trade-offs in run-time, space, and fidelity require the careful choice of simulation software. We used two software tools, one to characterize a process's behavior, the other to simulate the robots executing the appropriate composite controllers. The first simulates pseudo-dynamics during controller construction, while the second is a traditional sensor-based simulator used for testing and validation.

4.1 Microcanonical Metropolis Monte-Carlo (MMMC)

Microcanonical Metropolis Monte-Carlo (MMMC) is a numerical method described by Gross [2] as a variation of the Metropolis-Hastings algorithm. Both methods approximate a probability distribution by drawing a sequence of random samples.

Numerical methods are particularly useful when considering the thermodynamics of finite systems. The need for MMMC arose in nuclear physics because distributions over unconventional thermodynamic variables (conserved extensive variables) were considered. The resulting distributions are extremely peaked, with values ranging many orders of magnitude. Naïve sampling fails to obtain sufficient data for the lower probability states because they arise so infrequently. The MMMC algorithm decomposes the distribution surface into local patches called *knots*. Sampling then proceeds locally for each knot. Partial derivatives of the surface are calculated at each knot, and the function is then constructed through integration of those values.

Values at each knot are calculated by collecting statistics by a pseudo-dynamics simulation of the model from given initial value within the knot. The emphasis is on collecting numerical data based transitions within the knot itself. The simulation does not attempt to construct long temporal explorations of the states like a typical simulation would do. We believe that this approach, rather than simulation for plausible runs, is important for characterising the processes. From the perspective of the entropy surface, such an approach allows far more information to captured, which is necessary of a complete characterization of process behavior.

We used MMMC to construct the entropy surface for a finite Ising model. Even with 100 robots it is infeasible to explore all available states so a randomized algorithm must be used. Fig. 5 shows the logarithm of the probability normalized for each E value as calculated by MMMC. This function shows the extreme range of probabilities that can arise.

4.2 Microscopic simulation

Our sensor-based microscopic simulation uses an environment model in order to produce artificial readings for virtual sensors. We use an efficient Delaunay triangulation datastructure for representing positions of the robots as well as obstacles within the environment. Sensor readings are generated from this data structure rather than ray-casting in a bitmapped rendering. Robots are updated asynchronously.

The experiments used simulated robots with a velocity control interface. Linear and angular velocities were corrupted by three noise terms: multiplicative, additive and additive biased. The first two were drawn from normal distributions at each timestep, the last term represented a systematic bias and was drawn once at initialization. These values were different for each robot. The three terms for linear velocity were drawn from $\mathrm{N}(0, 0.01^2)$, $\mathrm{N}(0, 0.03^2)$, and $\mathrm{N}(0, 0.002^2)$; similarly for angular velocities $\mathrm{N}(0, 0.01^2)$, $\mathrm{N}(0, 2.5^2)$, $\mathrm{N}(0, 0.15^2)$; units are meters and degrees respectively.

The robots used three sensors: 1) a distance sensor with 12 radial rays, each with a range of 0.5 meters, with multiplicative $\mathrm{N}(0, 0.02^2)$, additive $\mathrm{N}(0, 0.05^2)$, and a additive bias for each ray $\mathrm{N}(0, 0.05^2)$; 2) a compass with four bits of information, and added noise $\mathrm{N}(0, 15.0^2)$ and added bias $\mathrm{N}(0, 2.0^2)$; 3) a single-bit sensor that responded to the sites of interest, that returned false-negatives with probability 0.15 and false-positives with probability 0.08. Only the compass provided global information. Distance readings returned from other robots and obstacles were indistinguishable.

The simulator included a model of local-broadcast communication. Each robot could send messages that may be heard within a disk of radius 2m centered at the sender. Messages arrived at a robot distance d from the sender with a probability given by $0.02d^2 - 0.18d + 1.0$. We also provided a local point-to-point network: robots could provide an intended recipient. These packets were locally broadcast within the same 2m disk, but automatically discarded by non-matching recipients. The sender was notified of a successful transmission.

4.3 Comparison of methods

Each run of the microscopic simulation produces a single temporal evolution of the robot system. It allows for comparison of task performance over time, with each independent execution offering new evidence. If the simulation is sufficiently realistic, it offers a forecast of performance with physical robots, and is thus important for validation purposes.

In contrast, MMMC does not simulate a single plausible trajectory. Instead, for each knot it simulates a brief temporal sequence and reinitializes to a value within the knot's parameter space. That brief temporal sequence may fail to follow the same strict dynamics rules as the robot system might. In the case of the Ising model, random flips are permitted, even if they are not neighboring. The key aspect is that the method provides a high-level characterization of the statistical aspects of the system.

Microscopic simulation is often used for iterative system design: insights from simulation (often supplemented by analysis) provide incremental improvements. Such methods operate in an end-to-end manner. The probabilistic characterization, like MMMC, provides a complementary approach.

5 Results

Experiments were performed in a $50\text{m}\times 50\text{m}$ arena with two disks of radius 3m placed at the North-East (NE) and South-West (SW) corners, 2m from the sides. The disks represent the sites of interest and can be sensed by robots that are positioned over them. Without localization information and equipped with a noisy 4-bit compass, many robots reach the arena corner having missed or failed to sense the disk. Independent sensing by the many robots within the swarm lessens the effects of the position and sensor uncertainty. Robots were initially placed in the arena center and were tasked with visiting first the NE site, then the SW site. The critical issue is in synchronization of the decision to advance from one site to the next.

Synchronization is achieved by coupling Processes 1 and 2 together as described above. On each robot the low-level controller (for obstacle avoidance, navigation, sensor processing, etc.) uses a variable to track task state. The value of this variable affects the interpretation of the compass readings and steering. The low-level controller is coupled to the two synchronization processes in two ways:

1. Input through gradual perturbation of the Process 1's state space. A robot detecting that it is over a site will increase the value of its Process 1 state (by 200 units). Observation of thrashing or lack of progress (by measuring odometry movement of less that 0.25m in 5s) results in the Process 1's state being reset to zero.
2. Output produced by monitoring average values of Process 2's the slow changing state variables. Values are averaged over a 12 second window. When the value is

within a threshold of zero (we used 0.02) a flag was set indicating that the task-state variable would soon change. When the mean approaches either -1 or 1 (we used 0.98), the task-state variable is altered to show the beginning of the next task (on -1) or return to the previous task (on 1).

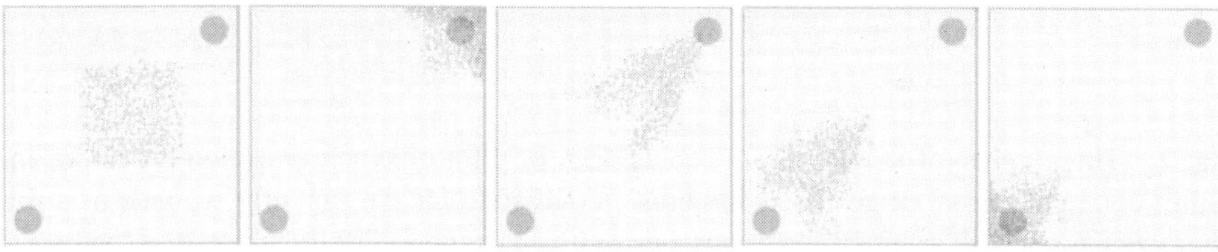

Fig. 6. Screen-shots of a simulation with 441 robots running for $\sim$ 520 seconds.

Fig. 6 shows simulation snap-shots of robots moving from the arena center to the first and second inspection sites. Fig. 7 gives connectivity information for the same run. The increasing density at each inspection location had a marked effect on the number of neighbors each robot had. Fig. 8 gives a plot of $\mathcal{K}_1$ and the average M value for three experimental runs. The figure shows Process 2's state being switched from -1 to $+1$ throughout the system. These three runs show similar behavior because the symmetry was broken in identical ways for each case. In 10 total runs, 4 cases transitioned back to -1 and the robots stayed at the NE site. This is expected as symmetry breaking occurred in an unbiased fashion; adding a bias would stop such cases from having to undergo the phase transition multiple times in order to reach the decision to explore the next site.

6 Discussion and Conclusion

The experiments show that the gross simplifications made in modelling the individual processes (especially Process 2) do not invalidate the general features of the processes behavior. For example, the mean connectivity is shown to be significantly higher than the value of 4 used in MMMC simulations. Also, the linearization arguments based on global constraints (e.g., in assuming fixed $\mathcal{K}_1$), statements about "slow" couplings, and ignorance of sparse node or network failures, are all only plausible to a limited extent. The experiments suggest that such assumptions are valid provided only very coarse features of the process behaviors are considered.

The predictions of process behavior, even with major modelling simplifications, can be useful provided the designer seeks a qualitative understanding of the behavior. The model of the synchronisation processes presented here is robust because it is based on the topological properties of the entropy surface. Our ongoing work is considering other processes that result in particular properties of the entropy surface.

Fig. 7 suggests that a smaller communication disk may suffice for the sequential inspection in the environment we considered. An alternative approach (perhaps without ergodic processes) could use the number of robots within communication range in order to switch behavior directly. An interesting question is whether such a switch would occur as abruptly as in our approach.

In conclusion, we have demonstrated that the composition of ergodic processes is a feasible approach for tackling the synthesis problem, at least in the case of simple

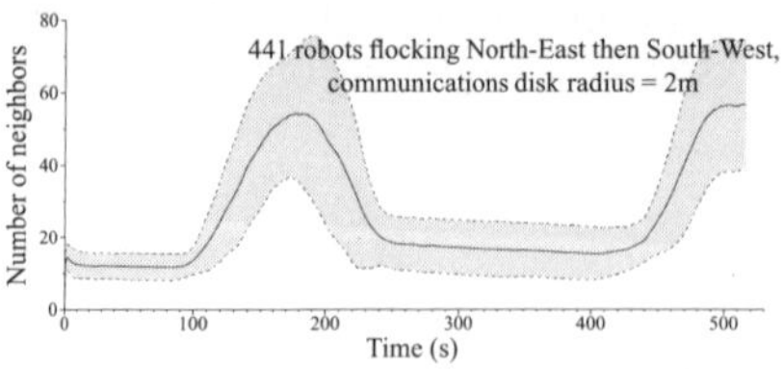

Fig. 7. The number of neighbouring robots within communication range for 441 robots. The average over all robots plus/minus one standard deviation.

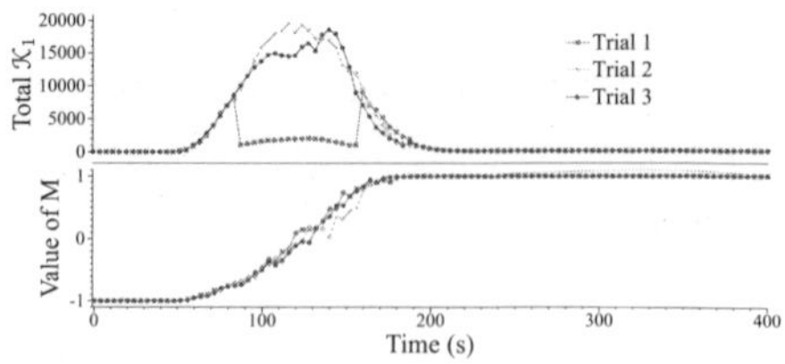

Fig. 8. Plot of synchronisation processes internal state for the first 400 seconds of a run with 441 robots. The transition of M values is clearly visible.

tasks. Synchronization, as a basis for sequencing, can be achieved by such processes. Composition of processes is useful because each can be independently analyzed in a manner that remains valid for combinations of ergodic processes. The analysis attempts to understand the processes' macroscopic behavior by considering the entropy surface. Numerical tools are useful for expanding the set of processes which are amenable to analysis, and, in particular, for dealing with issues that arise within finite systems. We view analysis of composable processes as complementary to traditional methods that iterate controller design at the complete system level. This approach appears relatively robust to modelling errors, provided the description is used for qualitative understanding of system behavior.

Acknowledgements This work was partially funded by National Science Foundation Grant No. IIS-0413321. We thank E. Mower for comments on an early draft.

References

1. M. E. Fisher. The theory of equilibrium critical phenomena. *Reports on Progress in Physics*, 30:615–730, 1967.
2. D.H.E. Gross. Microcanonical thermodynamics and statistical fragmentation of dissipative systems – the topological structure of the n-body phase space. *Physics Reports*, 279:119–202, 1997.
3. J. Jennings and C. Kirkwood-Watts. Distributed mobile robotics by the method of dynamic teams. In *DARS*, pages 46–56, Karlsruhe, Germany, May 1998.
4. C. V. Jones and M. J. Matarić. Automatic Synthesis of Communication-Based Coordinated Multi-Robot Systems. In *IEEE/RSJ IROS*, pages 381–387, Sendai, Japan, September 2004.
5. K. Lerman and A. Galstyan. Mathematical Model of Foraging in a Group of Robots: Effect of Interference. *Autonomous Robots*, 13(2):127–141, 2002.
6. A. Martinoli, K. Easton, and W. Agassounon. Modeling of Swarm Robotic Systems: A Case Study in Collaborative Distributed Manipulation. *IJRR*, 23(4):415–436, 2004.
7. L. E. Parker, M. Chandra, and F. Tang. Enabling Autonomous Sensor-Sharing for Tightly-Coupled Cooperative Tasks. In *Proceedings of the NRL Workshop on Multi-Robot Systems*, pages 119–230, Washington, DC, USA, March 2005.
8. K. Petersen. *Ergodic Theory*. Cambridge University Press, Cambridge, England, 1983.
9. P. White, V. Zykov, J. Bongard, and H. Lipson. Three dimensional stochastic reconfiguration of modular robots. In *Proceedings of Robotics: Science and Systems*, pages 161–168, Cambridge, MA, USA, June 2005.

A Study on Proportion Regulation Model for Multi-Robot System

Ken Sugawara[1] and Tsuyoshi Mizuguchi[2]

[1] Tohoku Gakuin University, 2-1-1, Tenjinzawa, Izumi, Sendai, 981-3193, Japan
sugawara@cs.tohoku-gakuin.ac.jp
[2] Osaka Prefecture University, 1-1, Gakuen-cho, Sakai, 599-8531, Japan
gutchi@ms.osakafu-u.ac.jp

Division of labor performed by social insects is one of the most advanced functions that they have evolved. For effective performance, the groups exhibit proportional regulation of population, which implies the proportion of each group is regulated against external disturbances without aid from a special individual. This paper addresses a variable probability model fot the proportion regulation of population in a homogenous multi-robot system, in which the state transition rate is combined with the external materials produced by corresponding task execution. Performance of the proposed model is confirmed by numerical simulations and experiments of simple multi-robot system.

1 Introduction

Cooperative multi-robot system is one of the most attractive topics in robotics, and many researchers have investigated multi-robot systems[1]. Some of them were inspired by biological systems and analyzed their performance using real robot systems[2, 3, 4, 5]. Actually, animals living in groups often show advanced functions or performances which exceed the simple sum of individual abilities. Especially, social insects like ants, bees, and termites exhibit some remarkable behaviors, such as colony formation, age polyethism, group foraging, etc[6, 7, 8]. The important point of their behavior is that their collective behaviors do not require a special individual which controls the behavior of the entire group. In spite of the lack of the special, behavior of the group is adaptable, flexible and robust against environmental disturbances. In their groups, hereditarily homogeneous individuals achieve these collective behaviors by interacting with each other through direct visual informations or chemical materials, such as pheromones.

Our interest in their characteristics is division of labor and proportion regulation of population. In their society, division of labor plays an important role, and it is known that proportion regulation of population are observed in division of labor for effective working. Fig.1 is a schematic figure of the "division of labor" and "proportion regulation of population" in the homogeneous multi-robot system that is treated here.

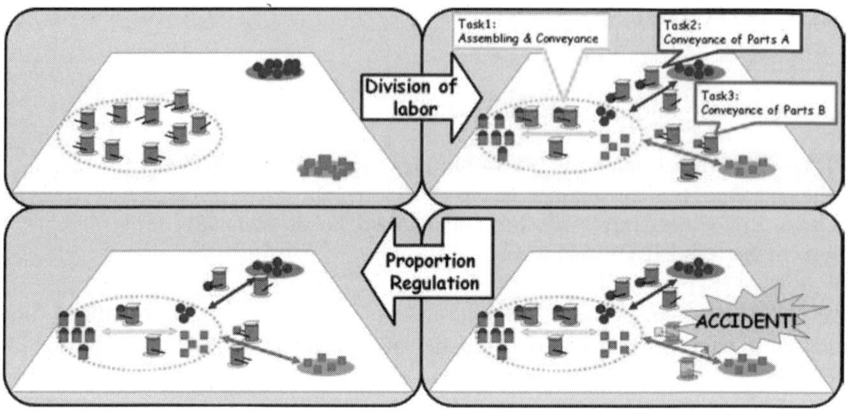

Fig. 1. A schematic figure of "division of labor" and "proportion regulation of population" in homogeneous multi-robot system.

Division of labor is also one of the most attractive topics in the study of multi-robot system[9, 10]. It will be important to consider division of labor and proportion regulation when the system needs to execute complex tasks autonomously and cooperatively. This paper addresses simple model of the proportion regulation of population in a homogenous multi-robot system, in which the state transition rate is combined with the external materials produced by corresponding task execution.

2 Model for division of labor and proportion regulation

As described above, division of labor can be widely observed in social insects. Many researchers, including biologists and theoretical scientists, have been interested in this phenomenon, and several mathematical models have been investigated.

Response threshold model

Bonabeau et al.[11] have developed a simple mathematical model based on response threshold. In this model, each individual has a response threshold for every task. They engage in a task when the internal stimuli exceeds the threshold of the task.

Based on this model, some experimental studies have been reported[4], in which the homogeneous group of robots divide into foraging robots and inactive robots effectively. This model functions well, if the threshold is scattered depending on the individual differences. However, it may cause "oscillatory" phenomena, when the scatter is small.

Non-linear dynamical model

A non-linear dynamical model[12] was proposed, especially focusing on the cell differentiation of Dictyostelium slug, in which all cells are divided into two types of cells, prestalk and prespore cells. The differentiation is flexible and it is known the ratio between them is almost constant. In this model, each element has two internal variables, - activator and inhibitor -, and they are coupled globally using their average quantities. The population "differentiates" into two states, and the system maintains the ratio against large disturbances by changing the state of the individuals.

Dynamic potential model

In a colony of social insects, we can observe "*age polyethism*," a division of labor based on individual age. In an insect colony, it is known that individuals change their tasks depending on their age. A dynamic potential model[13] consists of N indentical individuals, M tasks, M external materials and one-dimensional internal reference potential with m valleys. Each individual has an internal variable whose dynamics obey the effective potential. In this model, each individual detects the concentration of the external materials, and changes its internal variable depending on the effective potential derived from the concentration of the materials and the reference potential. Here, the proportion of population in each task can be controlled by the depth of the valleys in the potential. Since the reference potential is expressed by the valleys of the one-dimensional function, each individual changes its task sequentially. Hence, this model is suitable for expressing age polyethism.

Proposal

In this paper, we propose a variable probability model. Let us assume all the robots are homogeneous and each robot has some states and corresponding tasks. For simplicity, this paper treats the case that the robots have three states and three corresponding tasks. For the proportion regulation of population, each robot needs to determine its state appropriately. The simplest method to determine the proportion of the number of robots in state 1, 2, and 3 is to control the transition rates among them. Here the proportion can be expressed by the relative ratio of the transition rates. Let's denote the transition rates between the states as α_1, α_2, and α_3(Fig2(a)). If we set α_1, α_2, and α_3 as $0.1, 0.2$, and 0.3, respectively, the proportion of the number of robots in each state becomes $1 : 2 : 3$.

It should be mentioned that the proportion depends on the relative ratio at the equilibrium state and that the time to reach the equilibrium state depends on the transition rate itself. For example, if we compare the transition rate set $(\alpha_1, \alpha_2, \alpha_3) = (0.1, 0.2, 0.3)$ and $= (0.01, 0.02, 0.03)$, the proportion of population becomes 1:2:3 in both cases at the equilibrium state, but the balancing time of the former set is ten times faster than the latter one. It means that the balancing time can be independently controlled by w_j when we describe the transition rate as α_j/w_j instead of α_j. Fig.2(b) shows the schematic of the proposed model.

We also mentioned that we can generalize this model as shown in fig.2(c). It enables us to make an asymmetric transition flow,by describing the transition rate asymmetrically. It means that this model can also treat one-way transition such as age polyethism.

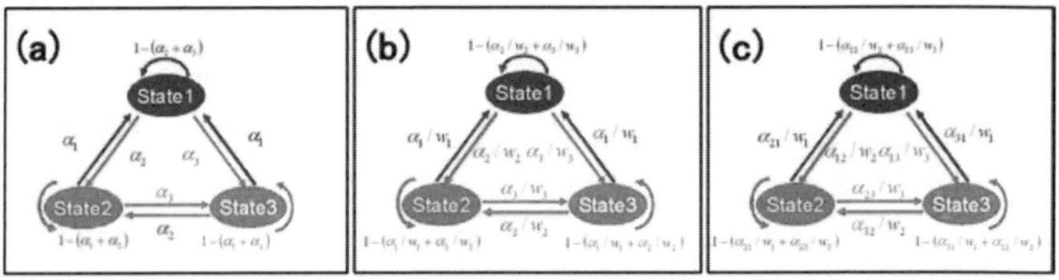

Fig. 2. State transition diagram. (a) Simple probability model. (b) Extended probability model. (c) Generalized transition model.

Next, we describe how to control the balancing time. When the system reaches the desired proportion, it is preferable that each robot does not change its state. It means w_j should be relatively large. On the contrary, when the balance of the proportion becomes worse, it is desirable to raise the transition rate in order to reach the desired proportion in a short time. It means w_j should be relatively small. Hence, it is necessary to introduce a dynamics to change the value of w_j flexibly depending on the balancing condition. In this paper, we realize the dynamics by combining the transition rates and the concentration of volatile external materials. Here we call this material as "a stock material", which is inspired by the chemical signals of social insects. As known well, social insects communicate by chemical signals and establish well-ordered society.
The quantity of the stock materials P_j is assumed to obey the following equation.

$$\dot{P}_j = -\tau_j P_j + c \cdot n_j,$$

where the first term on the right hand side represents the decay of the material, and the second represents the production of the materials by n_j robots which engage in task j. τ_j is a decay time of jth stock material. Here τ_j and c

is a constant. The value w_j, which regulates the transition rate, obeys the following equation.

$$w_j = \frac{k}{1 + exp(-\beta(P_j - L \cdot \alpha_j \cdot \sum P_m))},$$

where β, k, L are constant.

3 Simulation

In this section, we discuss the performance of the proposed model by numerical simulation. The simulation condition is as follows:
$n = 500, \alpha_1 = 0.5, \alpha_2 = 0.3, \alpha_3 = 0.2, \tau = 40.0, k = 20.0, L = 0.8$ and $\beta = 10.0$,
where the desirable ratio of the states 1, 2, and 3 is 5:3:2. As an initial condition, all the robots are assigned state 1 and the stock materials $w_j|_{t=0}$ are all set as zero.

Fig.3 shows a typical example of the time evolution of the simple probability model(a) and that of the proposed model(b). Responses to the disturbance applied to the proposed model are also shown in this figure, in which all robots in state 3 are forced to be state 1 (Fig.3(c)) and all robots in state 3 are removed (Fig.3(d)) at $t = 30000$.

Fig.3(a)(b) shows the average ratio between state 1, 2 and 3 is almost maintained as we designed in both models, but the fluctuation observed in the simple probability model is larger than that of the proposed model. Fig.4 shows the histogram of the number of robots in each state for a constant duration. As shown here, the variance of the proposed model is smaller than that of simple probability model.

In division of labor, it is desirable that the robot engages in the same task as long as possible. It is because task change often leads to the loss of cost, time, and so on. Adam Smith (1776) claimed that one of the most important points of the division of labor is "better organization of work, which saves time in changing task." Fig.5 shows the time evolution of the state in 100 robots which are selected randomly.

As shown here, the robots based on simple probability model change the task frequently, but the robots based on the proposed model tend to engage in the same task longer. Fig.6(a) is the time evolution of the total amount of task changes, and fig.6(b) is the frequency how long each robot engages We measured how often the robots changes the tasks and how long each robot engages in the same task (Fig.6).

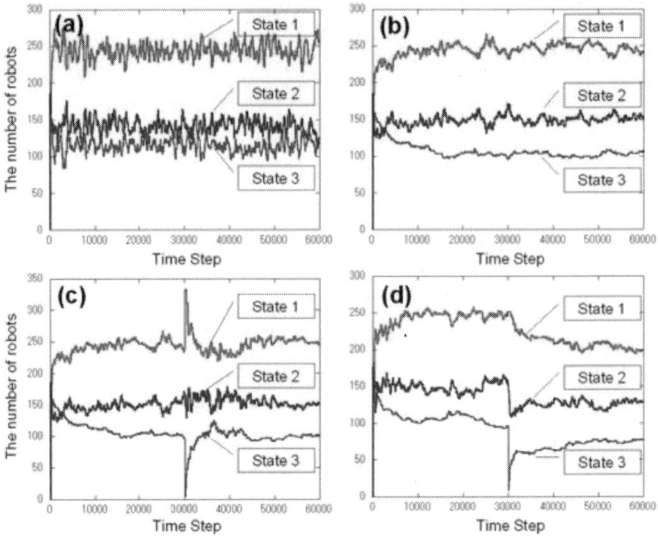

Fig. 3. The time evolution of the number of robots in each state. (a) Simple probability model. (b) Proposed model. (c) Response to the disturbance that all robots in state 3 is forced to be state 1. (d) Response to the disturbance that all robots in state 3 is removed.

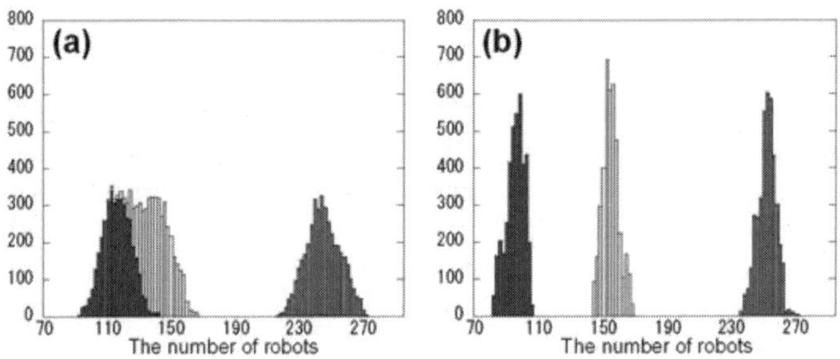

Fig. 4. Histogram of the number of robots in each state for constant duration. (a)Simple Probability Model. (b)Proposed model.

As you can see, the performance of the proposed method is much better than that of the simple probability model from the viewpoint of the task changes.

We also confirmed that the system shows same equivalent performance even if the transition rate is asymmetric. We set $\alpha_{31} = 0.5, \alpha_{12} = 0.3, \alpha_{23} = 0.2$ and the others are set as zero. One typical example is shown in Fig.7.

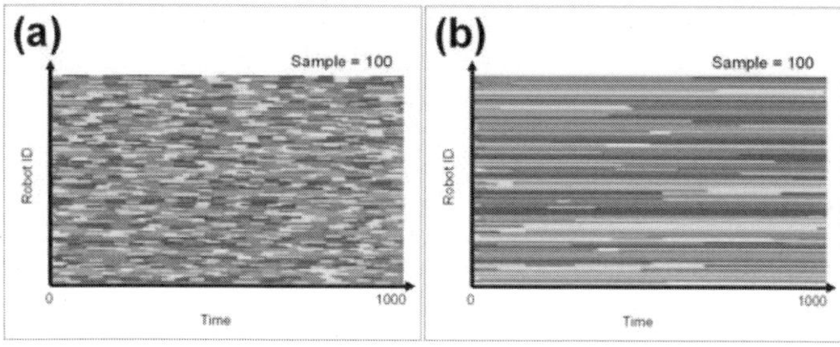

Fig. 5. Time evolution of the state in 100 robots. (a)Simple Probability Model. (b)Proposed model.

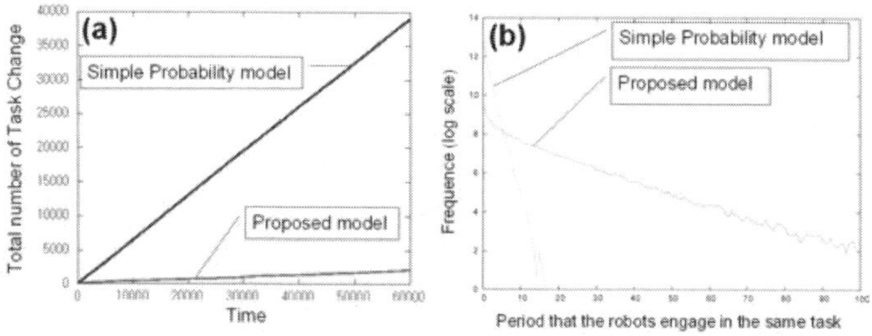

Fig. 6. (a) Total amount of the robots which change the task. (b) Frequency how long each robot engages in the same task.

In this figure, we confirmed the stability of the system. A ll robots in state 3 is forced to be in state 1 at $t = 30000$ as a large disturbance. As shown here, the system maintains the desired proportion even when large disturbance is added.

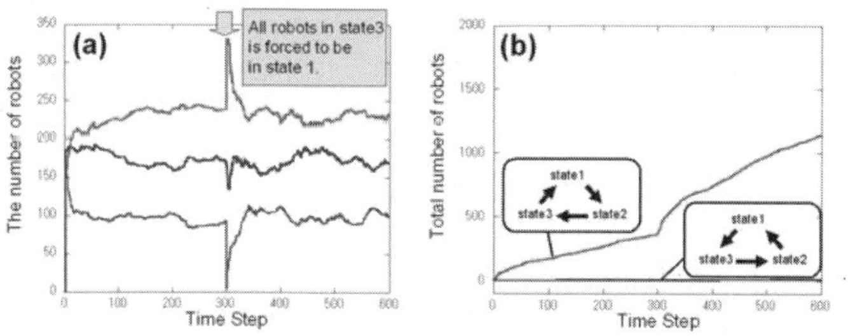

Fig. 7. Performance of the system which has an asymmetric state transition probabilities. All robots in state 3 is forced to be in state 1 at t=30000 as a large disturbance. (a) Time evolution of the number of robots in each task. (b) Total number of robots which change the task in each direction.

4 Experiment

4.1 Experimental equipment

First, we explain the overview of the experimental equipment. In the experiment, it is necessary to express volatile materials which can be detected by the robots. In this paper, we utilized "Virtual Dynamic Environment for Autonomous Robots (V-DEAR)[14]" for the robot experiment, in which the volatile materials are virtually expressed by light information (Fig.8(a)). This equipment comprises a LC projector to project the Computer Graphics, CCD camera to trace the position of the robots, and a PC to control them. Since the robots in the field have some light sensors on the top, they can measure the color and the brightness, as the field condition. Here we realize the dynamic interaction between the environment and robots.

In this experiment, we use a miniature robot "Khepera" (Fig.8(b)), which has a full-color LED and three color sensors on the top. Each robot indicates its condition by the LED, and acquires the light information using the sensors. The experimental field(Fig.8(c)) has three paths, which are assumed to be "task." Fig.8(d) is the schematic of the experiment. The robot in the nest measures the quantities of volatile materials, i.e. measures the color and the brightness of the light, and decides the task it should engage in. The field has three paths which represent "the tasks", and the robot passes the corresponding path. When it passes the path, the robot turns on the corresponding colored LED. V-DEAR detects the color of the LED on the top of the robot at the entrance and updates the quantity of the corresponding volatile material. Here the quantity of the materials is expressed by the brightness of the projected CG. Fig.8(e) is a snapshot during the experiment. Here the stock material 1, 2 and 3 are expressed by blue, red and green, respectively, and the quantities of the material are expressed by the brightness of each color.

4.2 Experimental result

We reveal the typical behavior of this system. Fig.9 is the time evolution of the number of robots in each state. Fig.9(a) and (b) are the time evolution of the number of robots in each state based on the simple probability model and on the proposed model. Here the desired ratio is 1:2:3. As the size of the group is small, fluctuation of the simple probability model seems to be large, but the proposed model maintains the desired ratio for long time. Fig.9(c) shows the response to large perturbation. Initial number of robots is 9, and initial population in each state is 3:3:3. Here the desired ratio is 1:1:1. We eliminated all the robots in state 1 at 120 s. As shown in this figure, the ratio converges to 1:1:1 rapidly by reallocation.

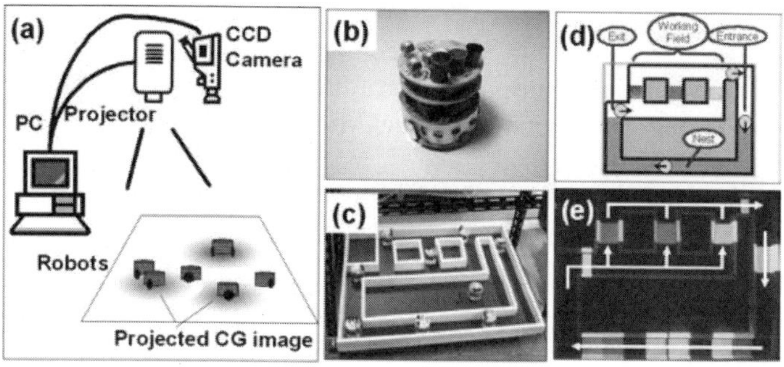

Fig. 8. (a) The schematic of V-DEAR. (b) Khepera robot for this experiment. Color sensors and a full-color LED are attached on the top. (c) The field for the experiment. (d) The schematic of the experimental field. (e) A snapshot during the experiment.

5 Conclusion

This paper addressed the proportion regulation of population in a homogenous multi-robot system. This was inspired by the task allocation of social insects. We focused on the proportion regulation of population and proposed the extended probability model. First, its performance was confirmed by numerical simulations. Second, we applied this model to a real robot system. Using V-DEAR and Khepera robots with extended sensors and indicators, we confirmed its effectiveness, especially by comparing it with the simple probability model.

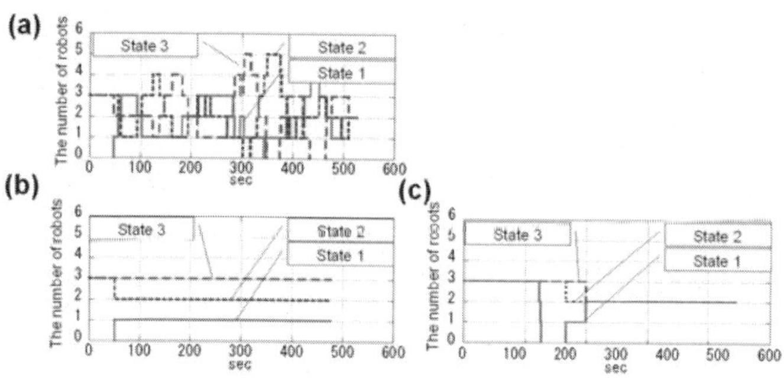

Fig. 9. Time evolution of the number of robots in each state. The total number of robots is six. The initial ratio is 0:1:1 and the target ratio is 1:2:3. (a) Simple probability model. As each robot frequently changes its task, the number of robots in each state fluctuates considerably. (b) Proposed model. The ratio rapidly converges to 1:2:3. (c) Response to large perturbation. In spite that all robots in state 1 are eliminated at 120s, the ratio converges to 1:1:1 rapidly by reallocation.

This work is partially supported by a Japanese Grand-in-Aid for Encouragement of Young Scientists from the Ministry of Education, Science and Culture (No.15760291.)

References

1. Y.U.Cao, A.S.Fukunaga and A.B.Kahng, "Cooperative Mobile Robotics:Antecedents and Directions," *Autonomous Robots*, 4, (1997) pp.7-27.
2. R. Beckers, O.E. Holland and J.L. Deneubourg, "From Local Actions To Global Tasks: Stigmergy and Collective Robotics," *Artificial Life IV*, MIT Press, (1994), pp. 181.
3. C. R. Kube and H. Zhang, "Collective Robotics: FromSocial Insects to Robots." *Adaptive Behavior* 2 (1994) pp.189-218.
4. M.J.B. Krieger, J.B. Billeter, "The call of duty: Self-organized task allocation in a population of up to twelve mobile robots", *Robotics and Autonomous Systems*, Vol.30 (2000) pp.65-84.
5. R. Beckers, O.E. Holland and J.L. Deneubourg, "From Local Actions To Global Tasks: Stigmergy and Collective Robotics," *Artificial Life IV*, MIT Press, (1994) pp. 181-189.
6. B. Hölldobler and E. O. Wilson, *JOURNEY TO THE ANTS*, Harvard University Press, (1994).
7. T. D. Seeley, *THE WISDOM OF THE HIVE*, Harvard University Press, (1995).
8. D. M. Gordon and M. Schwengel, *Ants at Work: How an Insect Society Is Organized*, Simon & Schuster, (1999).
9. C. Jones and M.J.Mataric, "Adaptive Division of Labor in Large-Scale Minimalist Multi-Robot Systems", *Proc. 2003 IEEE/RSJ Int. Conf. on Intelligent Robots and Systems (IROS-03)*, (2003) pp.1969-1974.
10. T.Balch and L.E.Parker ed. *Robot Teams: From Diversity to Polymorphism*, A K Peters, (2002).
11. E.Bonabeau, G.Theraulaz, and J.-L.Deneubourg, "Quantitative Study of the Fixed Threshold Model for the Regulation of Division of Labour in Insect Societies," *Proc. Roy. Soc. London B* 3 (1997) pp.191-209.
12. T. Mizuguchi, M. Sano, "Proportion Regulation of Biological Cells in Globally Coupled Nonlinear Systems",*Phys. Rev. Lett.*, Vol.75, No.5, (1995) pp.966-969.
13. K.Sugawara, T.Mizuguchi, H.Nishimori, "Collective Dynamics of Active Elements: group motions, task allocation and pheromone trailing, *Proc. Int. Symp. on Dynamical Systems Theory and Its Applications to Biology and Environmental Sciences*, (2003).
14. K. Sugawara, T. Kazama and T. Watanabe: "Foraging Behavior of Interacting Robots with Virtual Pheromone", Proc. 2004 IEEE/RSJ Int. Conf. on Intelligent Robots and Systems, (2004) pp. 3074-3079.

Market-Based Multi-robot Coalition Formation

Lovekesh Vig and Julie A. Adams

Vanderbilt University, Nashville TN 37212
{lovekesh.vig, julie.a.adams}@vanderbilt.edu

Summary. Task allocation is an issue that every multi-robot system must address. Recent task allocation solutions propose an auction based approach wherein robots bid for tasks based on cost functions for performing a task. This paper presents RACHNA, a novel architecture for multi-robot task allocation based on a modified algorithm for the winner determination problem in multi-unit combinatorial auctions. A more generic utility based framework is proposed to accommodate different types of tasks and task environments. Preliminary experiments yield promising results demonstrating the system's superiority over simple task allocation techniques.

1 Introduction and Motivation

Task allocation is a challenging problem due to the unpredictable nature of robot environments, sensor failure, robot failure, and dynamically changing task requirements. While market-based task allocation systems have traditionally found favor with the software-agent research community ([1], [2] and [3]), market-based control architectures are proving to be an effective distributed mechanism for multi-robot task allocation as well. Stentz and Dias [4] utilized a market-based scheme to coordinate multiple robots for cooperative task completion that introduced the application of market mechanisms to intra-team robot coordination. The common feature in market-based allocation mechanisms is an auction protocol to coordinate tasks between different robots [5], [6], [7] or between different components of the same robot [8], [9]. When an auction is announced, robots compute bids based on their expected profit for the tasks and the robots with the lowest cost bid are awarded contracts.

A number of elegant non market-based solutions to the task allocation problem have been proposed. The ALLIANCE [10] architecture uses motivational behaviors to monitor task progress and dynamically reallocate tasks. Recently Low et al. [11] proposed a swarm based approach for the cooperative observation of multiple moving targets (CMOMMT). Dahl et al. [12] present a task allocation scheme based on "Vacancy Chains," a social structure modeled on the creation and filling of vacancies in an organization. The Broadcast of Local Eligibility system (BLE) [13] system uses a Publish/Subscribe method to allocate tasks that are hierarchically distributed.

The common underlying factor in the above systems is the single robot-single task (SR-ST) assumption which entails that tasks are indivisible and may be performed by a single robot. As multi-robot tasks become more complex, this assumption is proving to be an oversimplification. Many task domains contain multiple tasks requiring a team of robots to work on them simultaneously, thus further complicating task allocation.

A relatively unexplored problem is the allocation of multi-robot teams to different tasks (the ST-MR problem), commonly known as the *Multi-Robot Coalition Formation (MRCF)* problem. Many coalition formation techniques within Distributed Artificial Intelligence (DAI) have been proposed for this provably hard problem [14], [15], [16]. Multi-robot coalition formation adds further complexity due to additional real world constraints [17] (fault tolerance, sensor location, communication costs, etc.). Recently a variety of market based solutions to the ST-MR task allocation problem have been proposed, [18], [19]. This paper proposes RACHNA[1], a novel market-based solution to the MRCF problem that leverages the inherent redundancy in sensor/actuator capabilities of robots to enable a more tractable, utility-based formulation of the MRCF problem.

This paper is organized as follows; Section 2 details the RACHNA architecture, the negotiation protocol, and the task environments. Section 3 provides the experimental details and results. Section 4 provides conclusions and outlines potential avenues for future work.

2 The RACHNA system

A common feature of the market based systems discussed in Section 1 is that they require the robots to bid on the tasks. The bidding process is central to determining the auction outcome. Therefore when dealing with complex tasks, the bidder should have a global view of the available resources. The RACHNA system reverses the bidding process. The auction is performed by the tasks for the individual robot services, thus allowing the bidding to be performed with a semi-global view of the resources necessary for coalition formation.

One of the most prominent differences between multi-agent and multi-robot domains is the level of redundancy in multi-robot and software-agent capabilities. Robots are manufactured on a large scale and are more likely to have greater redundancy in their sensor/actuator capabilities. RACHNA leverages this redundancy to enable a more tractable formulation of the MRCF

[1] Robot Allocation through Coalitions using Heterogeneous Non-Cooperative Agents

problem. RACHNA achieves this through the formulation of the MRCF as a multi-unit combinatorial auction. While single item auctions allow the bidders to bid on only one item, combinatorial auctions permit bidding on combinations of items.

Definition: The auctioneer has a set of items, M = *1, 2,..., m* to sell. The auctioneer has some number of each item available: $U = \{u_1, u_2, ..., u_m\}, u_i \epsilon Z+$. The buyers submit a set of bids, $B = \{B_1, B_2, ..., B_n\}$. A bid is a tuple $Bj = <(\gamma_j^1, ...,_j^m), p_j>$, where $\gamma_j^k \geq 0$ is the number of units of item k that the bid requests, and p_j is the price. The *Binary Multi-Unit Combinatorial Auction Winner Determination Problem (BMUCAWDP)* is to label the bids as winning or losing so as to maximize the auctioneers revenue under the constraint that each unit of an item can be allocated to at most one bidder:

$$max \ \sum p_j x_j \ s.t. \sum_{j=1}^{n} \gamma_j^i x_j \leq u_i, i = 1, 2, \ldots, m \tag{1}$$

The MRCF problem can be cast as a combinatorial auction with the bidders represented by the tasks, the items as the different types of robots, and the price as the utility that each task has to offer. Unfortunately, the BMUCAWDP problem is inapproximable [20] however some empirically strong algorithms exist [21], [20].

2.1 The Architecture

Two types of software agents are involved in the task allocation process:

1. **Service Agents** are the mediator agents through which the tasks must bid for a service. RACHNA requires that each robot have a set of services or roles it can perform. The roles are determined by the individual sensor and behavioral capabilities resident on each robot. One service agent exists for each service type that a robot can provide. A service agent may communicate with any robot that provides the particular service to which the agent corresponds. Service agents reside on any robot capable of providing the service. Thus, the global task information is acquired in a decentralized manner via service agents.
2. **Task agents** place offers on behalf of the tasks so as to acquire the necessary services. The task agents only communicate with the service agents during negotiations. Once the task is allocated, the task agent may communicate directly with the robots allocated to the task. Task agents may reside on a workstation or a robot and communicate with the necessary service agents.

An economy is proposed where the tasks are represented by task-agents that are bidding for the services of the individual robots. The economy has a

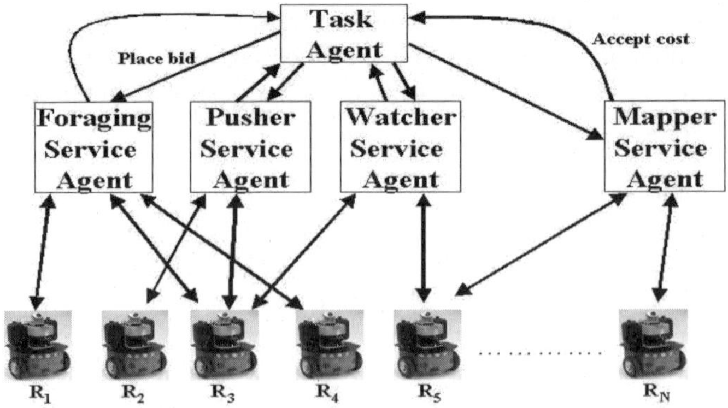

Fig. 1. An example RACHNA Implementation

set of robots $R_1, R_2, ..., R_N$ where each robot is equipped with sensor capabilities that enable it to perform various services such as pushing, watching, foraging, etc. The tasks are assumed to be decomposable into the sub-task behaviors. For example, a box-pushing task may require two pusher sub-task roles and one watcher sub-task role as shown in Figure 1. Each role is represented by a service agent that is responsible for negotiating with the robots that have the desired capability. The bids are relatively sparse compared to the overall space of coalitions and will yield a more tractable formulation of the MRCF. Unlike other heuristic based algorithms coalition formation [15], no restriction is placed on the coalition size.

2.2 The Allocation Environments

Three task types were permitted in the presented experiments:

1. **Urgent** tasks can pre-empt an ongoing standard task and generally have a higher average reward per robot. These tasks are emergency tasks that require immediate attention, such as fire extinguishing or rescue tasks.
2. **Standard** tasks are allocated only when sufficient free resources exist and when the task utility is sufficient to merit allocation. These tasks may be pre-empted by urgent tasks. Loosely coupled tasks (i.e. foraging) or tasks that may easily be resumed comprise this category.
3. **Non preemptable** tasks are allocated similar to standard tasks but cannot be pre-empted. Tightly coupled tasks fall into this category because, preemption would completely debilitate task performance.

Two different types of allocation are considered:

1. *Instantaneous Allocation*: A number of tasks are introduced into the environment and the algorithm allocates resources to the optimal set of tasks.

2. *Pre-emptive Allocation*: Involves introduction of a single urgent task that requires immediate attention. The urgent task offers higher rewards in an attempt to obtain bids from the robots.

Instantaneous Assignment

Instantaneous assignment requires multiple auctions while the system's objective is to allocate resources to tasks while maximizing overall utility. Services correspond to items, robots correspond to units of a particular item (service), and task offers correspond to bids. This work distributes this solution in order to leverage the inherent redundancy in robot capabilities, thereby obtaining a more tractable formulation of the MRCF problem.

The auction begins with each task agent sending request messages to the individual service agents. The service agents attempt to obtain the minimum possible price for the requested services. The robot's current minimum salaries are evaluated and a minimum increment is added in order to lure the robots to the new task. The service agents then forward this information to the task agents. The task agents determine if sufficient utility exists to purchase the required services. If this is the case, then the services are temporarily awarded the task. This offer-counteroffer process proceeds in a round robin fashion with the robots' salaries increasing at every step until there is a round where no service (robot) changes hands. At this point, a final stable solution is attained.

Random Assignment

Urgent tasks are randomly introduced and are allocated robot services according to a negotiation process between tasks. The negotiation begins when the new task submits a request to the required service agents for a certain number of services of that type. The service agents take into account the current robots' salaries and a bargaining process ensues with tasks increasing robot salaries until either the new task successfully purchases the resources or waits for additional free resources.

2.3 Utility vs. Cost

A difficulty with the employed market based approach is that the resulting teams are highly dependent on the initial utilities assigned to various tasks. However, this may not be an entirely undesirable property. While the system is sensitive to initial utilities, it also empowers the user to prioritize tasks by varying the task utilities. Most definitions of utility incorporate some notion of balance between quality and cost ([22], [23]). Cost quantification is relatively straightforward, however quantifying quality task execution prior to coalition formation for a new task can be difficult. Independent of the utility measure employed, what matters is that a mapping exists between coalition task pairs to scalar values permitting comparisons between coalitions for performing a task.

2.4 Multiple decompositions

Many scenarios involve more than one potential decomposition for a particular complex task and many possible decompositions may be considered when evaluating the potential coalitions. It may be possible to permit multiple decompositions via a task decomposition system [23] and introducing 'dummy' items to incorporate these, as described in [21].

3 Experiments

Preliminary experiments were conducted by simulating the RACHNA system on a single computer. The experiments recorded the variation in robot salaries and overall utility with bid numbers. A set of real world tasks were simulated in the Player/Stage environment to demonstrate task preemption.

3.1 Wage increase

The first set of experiments simulated a set of 68 robots and ten services such that each service had exactly ten possible robots capable of providing that particular service. 100 tasks were generated with each task requiring a random vector of resources. The variation in the average salary for each service type was recorded as the bids increased. Fig 2 shows the average, maximum, and minimum salary curves for all services. The results depict how the increasing competition (more tasks) increased the salaries as robots received better offers when demand increases. Initially the salaries are low (Number of tasks $\leq$ 20), the salaries rise at different rates depending on demand for a particular service ($20 \leq$ Number of tasks ≤ 40), and eventually if the demand for each service increases sufficiently, the salaries for all service agents approach high values (Number of tasks $=$ 100). The robots in RACHNA that are capable of performing services that are in high demand have a high likelihood of participating in the final allocation.

3.2 Effect of diversity

RACHNA leverages the redundant sensory capabilities in a set of robots in order to group robots and make the allocation problem more tractable. RACHNA does make any assumptions about the diversity of the resulting teams, only the diversity of the entire collection of robots. Fig 3 shows RACHNA's performance deteriorates as the number of services is increased and the number of robots remains constant. The higher the number of services, the lower the redundancy, and hence the higher the execution time of the algorithm. If there was only one service agent, all robots would be identical and task allocation is the least expensive. However, if each robot was different, task allocation would be more expensive. Thus the execution time increases with the increased diversity of the set of robots.

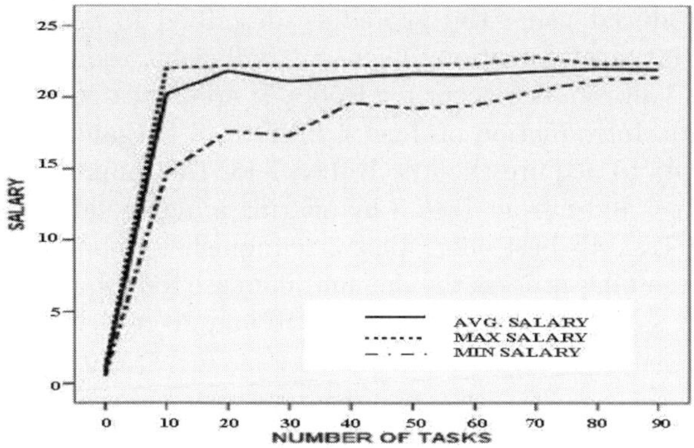

Fig. 2. Average salary across all robots vs. Tasks (bids).

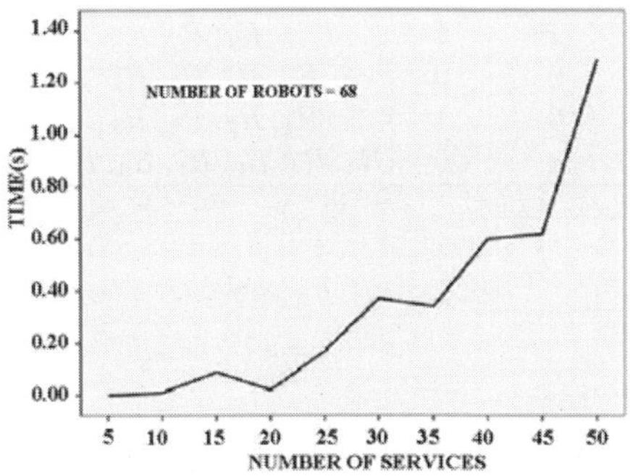

Fig. 3. Execution time vs. Number of Services.

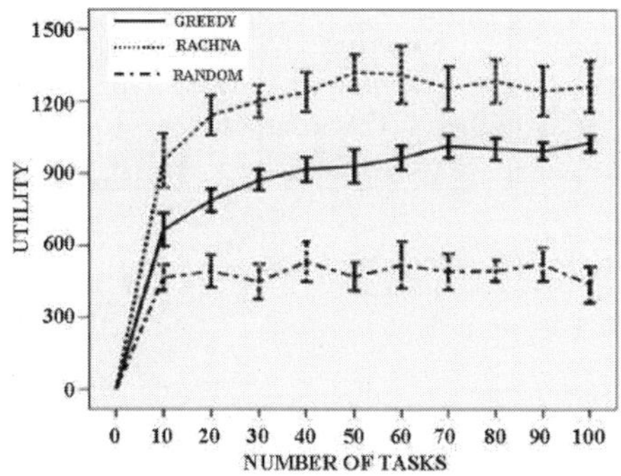

Fig. 4. Comparison of global greedy, random a allocations to RACHNA.

3.3 Utility Comparision

RACHNA's solution quality was compared to that obtained by two simple task allocation schemes. Fig 4 provides comparison for the average solution quality between solutions produced by RACHNA to those produced by the global greedy (best task first) and random allocation algorithms as the number of tasks varied. Each data point represents the mean performance from ten trials. A random task was generated for each trial and each algorithm's performance was recorded. RACHNA outperforms both the greedy and random allocation algorithms (as shown in Fig 4) because unlike greedy or random search, RACHNA refines the solution in each auction round to include better tasks (bids) and remove less profitable tasks.

3.4 Preemption Simulations

The preemption experiments involve a set of five services, ten robots, (see Table 1) and four heterogeneous tasks as described in Table 2. The first three tasks are introduced using the procedure described in Section 2. Fig 5(a) shows the initial task allocation. There are sufficient resources to satisfy all three tasks and all robots, except for Robot 7, are allocated with a minimum wage of 5 units. Introduction of Task 4 initiates a bargaining process where Task 4 attempts to acquire the idle Robot 7 for the minimum wage (5) and acquire Robots 6 and 8 from Task 3 by offering a higher salary. Since, Task 3 cannot match Task 4's best offer, Task 3 relinquishes Robots 6 and 8. At the end of this bargaining process the demand increase for robots of type 6 and 8 results in their salaries being increased to 10 and 20 respectively and Task 3 is preempted.

Table 1. Services

Services	Capabilities					Robots
	LRF	Camera	Bumper	Gripper	Sonar	
Foraging	0	1	0	1	1	R_1, R_2
Pushing	1	0	1	0	0	R_3, R_4, R_6, R_7
Object Tracking	0	1	0	0	1	R_1, R_2, R_5, R_8
Sentry-Duty	1	0	0	0	0	$R_3, R_4, R_6, R_7, R_9, R_{10}$

Table 2. Tasks

Tasks	Services				Priority	Utility
	Foraging	Pushing	Object-Tracking	Sentry-Duty		
1	2	0	0	1	Standard	70
2	0	2	0	1	Non-premptible	40
3	0	1	2	0	Standard	45
4	0	2	1	0	Urgent	50

The results reported in this section demonstrate the potential applicability of the system to different types of tasks and environments. RACHNA also allows for a more generic, task-independent system. It is important to note the favorable comparison of the suggested allocation to simple techniques like global greedy and random allocation. The fact that the algorithm leverages sensor redundancy makes the coalition formation tractable and the experiments demonstrate the improved performance.

4 Conclusions and future work

This paper presents RACHNA, a market-based distributed task allocation scheme based on the multi-unit combinatorial auction problem. RACHNA

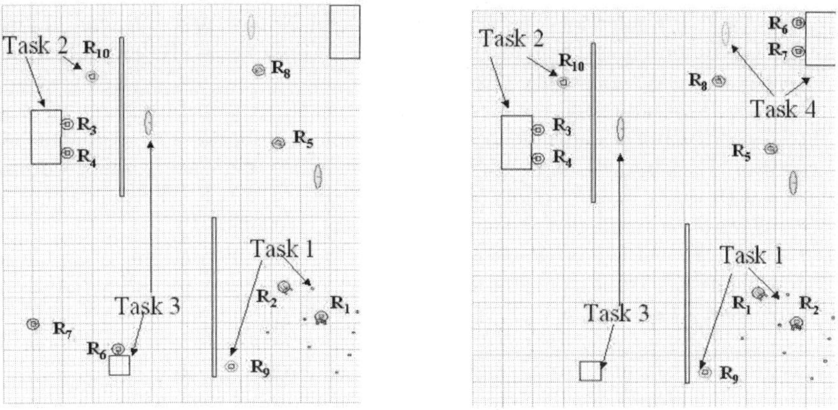

(a) The initial allocation of tasks 1,2 and 3.

(b) Allocation of task 4 and pre-emption of task 3.

Fig. 5. The pre-emption of the standard task 3 by the urgent task 4

reverses the auction scheme found in other market-based coordination schemes by allowing tasks to bid on robot services rather than the other way around. RACHNA is a utility based system, allowing the user to specify the task priority. Finally the system produces higher quality solutions than simple greedy or random task allocation strategies and enables a more tractable formulation of the coalition formation problem by leveraging the redundancy in robot sensor capabilities. Future work involves real robot experiments to demonstrate allocation and dynamic task preemption.

References

1. Sandholm, T.: An implementation of the contract net protocol based on marginal cost calculations. In: Proceedings of the Eleventh National Conference on Artificial Intelligence. (1993) 256–262
2. Collins, J., Jamison, S., Mobasher, B., Gini, M.: A market architecture for multi-agent contracting. Technical Report 97-15, University of Minnesota, Dept. of Computer Science (1997)
3. Sycara, K., Zeng, D.: Coordination of multiple intelligent software agents. International Journal of Intelligent and Cooperative Information Systems **5** (1996) 181–211
4. Stentz, A., Dias, M.B.: A free market architecture for coordinating multiple robots. Technical Report CMU-RI-TR-01-26, Robotics Institute, Carnegie Mellon University (1999)
5. Gerkey, B.P., Matarić, M.J.: Murdoch: Publish/subscribe task allocation for heterogeneous agents. In: Proceedings of Autonomous Agents. (2000) 203 – 204
6. Botelho, S.C., Alami, R.: M+: A scheme for multi-robot cooperation through negotiated task allocation and achievement. In: Proceedings of IEEE International Conference on Robotics and Automation. (1999) 1234 – 1238

7. Dias, M.B.: TraderBots: A New Paradigm for Robust and Efficient Multirobot Coordination in Dynamic Environments. PhD thesis, Robotics Institute, Carnegie Mellon University (2004)
8. Laengle, T., Lueth, T.C., Rembold, U., Woern, H.: A distributed control architecture for autonomous mobile robots. Advanced Robotics **12** (1998) 411–431
9. Caloud, P., Choi, W., Latombe, J.C., Pape, C.L., Yim, M.: Indoor automation with many mobile robots. In: Proceedings of the IEEE/RSJ International Conference on Intelligent Robots and Systems. (1990) 67–72
10. Parker, L.E.: ALLIANCE: An architecture for fault tolerant multi-robot cooperation. IEEE Transactions on Robotics and Automation **14** (1998) 220–240
11. Low, K.H., Leow, W.K., M. H. Ang, J.: Task allocation via self-organizing swarm coalitions in distributed mobile sensor network. In: Proceedings of the American Association of Artificial Intelligence. (2004) 28–33
12. Dahl, T.S., Matarić, M.J., Sukhatme, G.S.: Multi-robot task-allocation through vacancy chains. In: Proceedings of IEEE International Conference on Robotics and Automation. (2003) 14–19
13. Werger, B., Matarić, M.J.: Broadcast of local eligibility: Behavior based control for strongly cooperative multi-robot teams. In: Proceedings of Autonomous Agents. (2000) 21–22
14. Sandholm, T.W., Larson, K., Andersson, M., Shehory, O., Tomhe, F.: Coalition structure generation with worst case guarantees. Artificial Intelligence **111** (1999) 209–238
15. Shehory, O., Kraus, S.: Methods for task allocation via agent coalition formation. Artificial Intelligence Journal **101** (1998) 165–200
16. Abdallah, S., Lesser, V.: Organization-Based Cooperative Coalition Formation. In: Proceedings of the IEEE/WIC/ACM International Conference on Intelligent Agent Techonology. (2004) 162–168
17. Vig, L., Adams, J.A.: Issues in multi-robot coalition formation. In: Proceedings of Multi-Robot Systems. From Swarms to Intelligent Automata. Volume 3. (2005) 15–26
18. Zlot, R., Stentz, A.: Complex task allocation for multiple robots. In: Proceedings of the IEEE Conference on Robotics and Automation. (2005) 67–72
19. Lin, L., Zheng, Z.: Combinatorial bids based multi-robot task allocation. In: Proceedings of the International Conference on Robotics and Automation. (2005) 1145–1150
20. Sandholm, T.: Algorithm for optimal winner determination algorithm. Artificial Intelligence **135** (1998) 1–54
21. Leyton-Brown, K., Y. Shoham, M.T.: An algorithm for multi-unit combinatorial auctions. In: Proceedings of the 17th National Conference on Artificial Intelligence. (2000) 56–61
22. Gerkey, B., Matarić, M.J.: A framework for studying multi-robot task allocation. In: Proceedings of the Multi-Robot Systems Workshop: From Swarms to Intelligent Automata. Volume 2. (2003) 15–26
23. Tang, F., Parker, L.E.: ASyMTRe: Automated synthesis of multi-robot task solutions through software reconfiguration. In: Proceedings of the IEEE International Conference on Robotics and Automation. (2005) 1501–1508

Multi-robot User Interface Modeling

Alan R. Wagner, Yoichiro Endo, Patrick Ulam, Ronald C. Arkin

Mobile Robot Lab
College of Computing
Georgia Institute of Technology
Atlanta, Georgia 30332-0250
{alan.wagner, endo, pulam, arkin}@cc.gatech.edu

Abstract. This paper investigates the problem of user interface design and evaluation for autonomous teams of heterogeneous mobile robots. We explore an operator modeling approach to multi-robot user interface evaluation. Specifically the authors generated GOMS models, a type of user model, to investigate potential interface problems and to guide the interface development process. Results indicate that our interface design changes improve the usability of multi-robot mission generation substantially. We conclude that modeling techniques such as GOMS can play an important role in robotic interface development. Moreover, this research indicates that these techniques can be performed in an inexpensive and timely manner, potentially reducing the need for costly and demanding usability studies.

Keywords. User interface design, Goals Operators Methods Selection Rules (GOMS), Contract Network Protocol (CNP), Case-Based Reasoning (CBR)

1 Introduction

This paper investigates the problem of user interface design and evaluation for teams of heterogeneous, cooperative, and (generally) autonomous mobile robots. As part of NAVAIR's Intelligent Autonomy program, the purpose of this research is to reduce the burden of deploying teams of heterogeneous robots and conducting multi-robot missions. The methods we detail, however, are general purpose and applicable to virtually any software interface. Moreover, we contend that these methods hold special promise for developers of multi-robot systems.

Our approach is motivated by the notable challenges multi-robot systems present to user interface designers. Currently, as the size of the multi-robot team increases, so do the startup, running, and maintenance demands placed on the user [1]. Users can become overwhelmed with situation awareness information in high workload environments. Moreover, because multi-robot applications tend towards mission critical domains, such as search and rescue and military domains, users must be capable of quickly and easily assessing important situation information while also being shielded from insignificant or redundant information.

This paper details a modeling approach to multi-robot user interface evaluation. This approach is used to assess the effectiveness of novel interface and algorithmic changes made to an existing multi-robot software toolset. Roboticists have traditionally conducted usability experiments to gauge the state of their interface designs [2]. Usability experiments, however, are expensive, time-consuming, and too cumbersome for effective interface development [3]. Moreover, for specialized software, such as multi-robot mission specification systems, usability testing typically requires difficult to find domain experts to assess the interface design. Alternatively, random or semi-random subject populations are used. These populations, however, tend to have little or no experience with the application in question, whereas, in reality, the target population may have significant experience with either previous versions or similar types of software. These reasons serve as a motivation to explore alternatives to formal usability studies such as GOMS (Goals, Operators, Methods and Selection rules) modeling [4].

2 Related Work

At least two aspects of user interface design differentiate existing evaluations in robotics from those in software engineering and HCI. First, within robotics researchers typically evaluate user interface designs at the system architecture level (e.g., [5]), rather than at the level of particular algorithms and features. Some exceptions exist [2]. Software engineering practices and Human Computer Interaction (HCI) researchers, on the other hand, routinely examine the user interface implications of specific changes to existing software packages (e.g., [6]) as well as conducting system level evaluations. Second, usability studies represent the majority of formal user interface evaluations within robotics [2]. The software engineering and HCI disciplines, on the other hand, employ a wider array of user interface techniques including heuristics and user modeling [5]. A GOMS model is a type of user model that describes the knowledge a user must posses in order to perform tasks with the system [3]. Research by Yanco et al. represents the only example of GOMS based user interface design evaluation within robotics located to date [7]. Yanco et al. focus on the challenge of a specific robotics domain and a usability coding scheme inspired by GOMS. Our intention, rather, is to explore the use of GOMS as a primary means of evaluating incremental additions to a multi-robot user interface.

3 User Interface Modeling

A GOMS model explicitly represents the knowledge that a user must have in order to accomplish goals using an interface [3]. Natural GOMS Language (NGOMSL) is one method for explicitly representing GOMS models [8]. NGOMSL is a structured language notation in program form (see examples below). As a knowledge representation, a GOMS model can also serve to characterize ongoing user decisions or as a description of what a user must learn. Moreover, because user goals tend to be constrained by interface design, GOMS models can quantitatively predict aspects of usability such as the efficiency and simplicity of procedures.

A user interface analyst conducts a GOMS analysis by a describing in detail the goals, operators, methods, and selection rules a user must follow for a set of tasks. A goal is something that a user tries to accomplish. For example, one goal resulting from our GOMS analysis is to edit the parameters of a robot behavior. An operator is an action that a user executes. An example of an operator is moving the cursor to a screen location. A method is a sequence of operators for accomplishing a goal. The following example method (in NGOMSL) accomplishes the goal of adding a mission behavior:

```
Method for goal: add behavior                 KLM op  Time(s)
  Step 1: Locate add behavior icon              M       1.2
  Step 2: Move cursor to add behavior icon      P       1.1
  Step 3: Click mouse button                    BB      0.2
  Step 4: Think-of new icon location            M       1.2
  Step 5: Move cursor to new location           P       1.1
  Step 6: Click mouse button                    BB      0.2
  Step 7: Return with goal accomplished       Total     3.0
```

Finally, a selection rule (also in NGOMSL) routes control to the correct method for accomplishing a goal when many possible methods are possible.

A GOMS analysis begins by first describing a top-level goal and its associated high-level operators and by then iteratively replacing these operators in a breadth-first manner with methods and selection rules that accomplish each goal until all of the operators are primitive and cannot be further analyzed. The analyst may choose his or her own primitive operators, but typically, standard primitive operators from the Keystroke-Level Model (KLM) are used [3]. These primitives offer a well-documented mean time of operation and in some cases functional estimates.

Once the GOMS analysis is complete, the analyst can perform a qualitative evaluation of the interface to examine the efficiency, consistency, and cleanliness of the design. The analyst can also use the GOMS model to estimate the execution time of specific user tasks and the amount of effort that it will take users to learn procedures represented in the model.

GOMS modeling, however, is not without limitations. The execution and efficiency predictions generated from a GOMS model assume error-free performance. A GOMS model therefore represents a best-case evaluation of an interface. Still, GOMS models provide a valuable baseline for comparison of interface changes. GOMS modeling can also require subjective decisions and judgment calls. In spite of these subjective decisions, the analyst objectively constructs the majority of the model based on the actual state of the interface design. Overall, GOMS modeling serves more to guide interface development than to completely replace usability testing [6].

4 A Case Study in Multi-robot User Interface Modeling

GOMS modeling has been successfully employed within HCI and software engineering [7, 9], but is relatively unknown within robotics [7]. It is our contention that GOMS assessments of multi-robot user interface designs could play a vital role in the generation and rapid prototyping of future multi-robot system interfaces. To explore this hypothesis we conducted a detailed GOMS analysis of features recently added to the *MissionLab* toolset [9].

MissionLab allows users to generate multi-robot missions in the form of a FSA (Finite State Acceptor) in which nodes representing the robot's behaviors are connected via directed edges representing the robot's perceptual trigger schemas. The FSA serves as a flexible robot mission and can be stored, copied, or edited as needed to generate novel missions. This software system also features a Case-Based Reasoning (CBR) wizard that abstracts entire multi-robot missions as cases to be matched to the user's needs (Figure 1 left) [2]. The CBR wizard can also use cases as high-level drag-and-drop robot tasks, hence simplifying the mission creation procedure.

Our current investigation considers a scenario where multiple, heterogeneous robots are available for tasking. With respect to mission generation, multi-robot tasking presents additional challenges to the user. In a system of many robots, or when each robot affords unique capabilities, the user may have to assign each task to a robot. The generation of a multi-robot mission, in this case, demands (1) the user delineate the tasks necessary for the mission, and (2) the user assign each of these tasks to a specific robot or robots. The CBR wizard eases the first challenge but does not assist with the second.

To manage these challenges we have developed a novel method for generating multi-robot missions which employs a Contract Net Protocol (CNP) working in conjunction with the CBR Wizard to reduce the burdens placed on the user (see [10] for a review of multi-robot CNP). In its most general form, CNP is an auction-style algorithm in which the robots of a multi-robot system produce bids based on their estimate of their ability to perform the auctioned task. Typically, when the auction closes CNP assigns the highest bidder the task. Our system uses CNP as a method to aid the user by assigning robots to specific tasks prior to the start of the mission. In this role, the goal of the pre-mission CNP system is thus to generate an a priori mapping of robots to available pre-mission tasks.

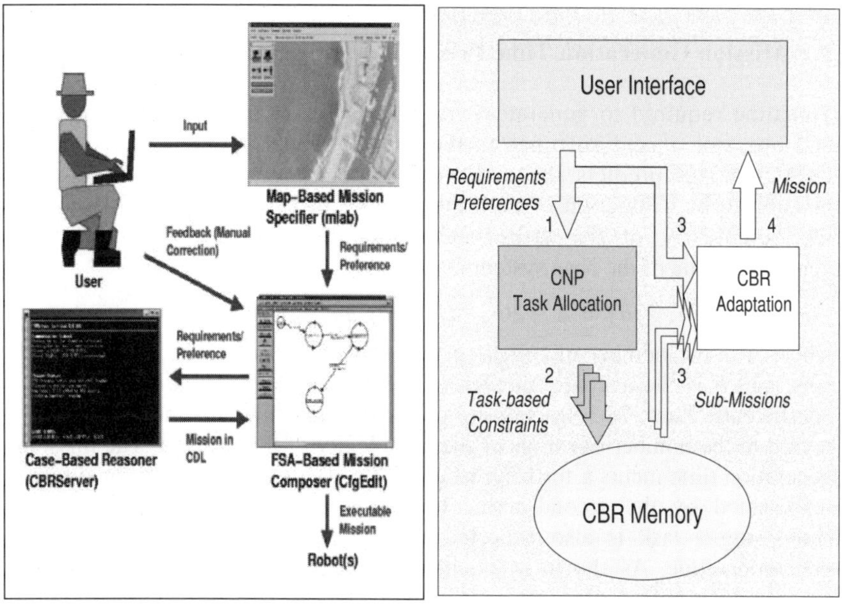

Fig. 1. Integrated CBR-CNP system for multi-robot mission generation

This system operates by first relaying a set of robot requirements and task requirements to a CNP task allocation component (Figure 1 right). The CNP component then conducts an auction resulting in a robot-to-task mapping. The system then uses this mapping to retrieve a sub-mission from CBR memory for each robot. Using additional user task preferences, these sub-missions are adapted into a single full mission. Finally, the system presents the complete mission with robot-to-task assignments to the user for acceptance, alteration, or rejection.

To assess the usability of the new features the authors created two GOMS models, one for the base system (no CBR or CNP) and one for the integrated CBR-CNP system. Both models evaluate the general methods available to users for creating multi-robot multi-task missions, beginning with the decision to build a new mission and ending with a complete mission. In some cases, the analyst made judgment calls concerning which GOMS operator to use or execution time estimates. As much as possible the analyst strived to maintain consistency across both models. Each model required approximately 20-30 hours to construct and was created by the lead author using guidelines available from [3]. Additional supplementary data and both complete models are available at www.cc.gatech.edu/ai/robot-lab/onraofnc/data/goms-2005/.

5 Results

5.1 Mission Generation Time Predictions

The time required to generate a mission is determined from the method and operator execution times in the GOMS model [3]. The base system GOMS model predicts the generation time of a multi-robot multi-task mission to be a function of both the number of robots in the mission and the complexity of the tasks each robot is to perform. The mission generation time of the base system in seconds, t_{gb}, is predicted to be:

$$t_{gb}(n, g, h) = A + Bn + Cgn + Chn \qquad (1)$$

where n is the number of robots, g is the average number of behaviors per task, and h is the average number of triggers per task. Table 1 lists model coefficients. Thus, 78.3 seconds are required to generate any mission without regard to the number of robots or the complexity of the mission. The mission generation time incurs a further cost of 26.7 seconds for each additional robot represented by the second term. The number of behaviors and triggers composing a task is also expected to have large impact on the mission generation time. As shown by equation (1) 34.8 seconds are necessary per robot and per behavior or trigger. Alternatively, one can estimate the mission generation cost as:

$$t_{gb}(n, \tau) = A + Bn + 2Cn\tau \qquad (2)$$

where τ is the number of tasks. Equation (2) assumes that all tasks require the same number of behaviors and triggers ($2\tau = g + h$) and that a single task is equivalent to a single behavior and a trigger. This, however, is generally not the case and equation (2) is offered solely for comparison to the integrated CBR-CNP GOMS model.

Table 1. Model coefficient values.

	Initial Model Coefficient Values	Refined Model Coefficient Values
A	78.30	74.22
B	26.70	18.28
C	34.80	32.08
D	85.30	77.92
E	24.30	19.56

The integrated CBR-CNP GOMS model, on the other hand, predicts the generation time of a multi-robot multi-task mission will only be a function of the number of tasks. In this case, the mission generation time in seconds, t_{gi}, is governed by:

$$t_{gi}(\tau) = D + E\tau \tag{3}$$

where τ is the number of tasks. The integrated CBR-CNP model incurs a 7.0 seconds greater startup cost ($D - A$). This cost is primarily due to the need to select the option for CBR-CNP. Users incur a further cost for each task. Because the integrated system abstracts from the user the assignment of each robot to a task, mission generation is independent of the number of robots. Moreover, comparing equations (2) and (3), we note that the models predict that the integrated CBR-CNP system requires approximately 44.3 seconds less per task ($2C - E$; assumes a single robot) than the base system given that the assumptions mentioned above hold.

5.2 Learning Effort Predictions

The effort required to learn how to generate a mission is determined from the number and length of the methods in the GOMS model [3]. User learning effort is estimated from the number of NGOMSL statements in each model. The total number of NGOMSL statements in the model describes the amount of procedural knowledge a user must have in order to use all aspects of the software system. User training describes the process of learning this procedural knowledge. Hence, models with fewer individual statements require less effort to learn.

The GOMS model of the base system (no CBR or CNP) included 187 individual statements that encompassed the procedures necessary for creating a multi-robot multi-task mission. The GOMS model of the integrated CBR-CNP system, in contrast, included 147 operators. We, therefore, expect the base system to require approximately 21.3 % more procedural knowledge.

We did not conduct experiments to confirm this result. However, if one assumes that additional procedural knowledge results in less accuracy, then this result corroborates related prior usability studies conducted by our lab [2]. This earlier work examined the use of the CBR wizard for a variety of mission generation tasks and found that it improved the accuracy of mission generation on tasks requiring two robots by approximately 33%. Our GOMS models indicate that this increase in accuracy may partially result from the reduced workload on the user. Our models also predict that less accuracy will be gained when generating a single robot mission compared to a multi-robot mission. This was also found to be the case in [2].

5.3 Comparison of Model Estimations to Actual Expert Performance

The results from the previous two sections clearly and quantitatively indicate the value of the CBR-CNP. As far as the case study is concerned, these results are sufficient. We decided, however, to also investigate the methodology itself by examining the accuracy of the predicted execution times for both GOMS models. In particular, we hoped to determine (1) if the primitive KLM operators used for the models accurately reflected experimental values for expert users and (2) if these primitive operators are immune to experimenter bias. To accomplish this, we conducted an experiment involving system experts. These experiments attempted to gauge the accuracy of both the overall models and of several GOMS methods that could then use to refine the models. The experiment required the expert to create multi-robot multi-task missions 20 times using both the base system and the integrated CBR-CNP system. During data collection, time data was recorded related to all of the users' actions. The experts used for the study consisted of six members of the same research lab including three authors of this paper. We hypothesized that because GOMS operators consist of low-level primitives such as individual key strokes that experimenter bias would be minimal. Moreover, the authors decided which GOMS methods to compare after experimentation but before analyzing the data. Thus, no subject knew which part of the experiment would be used.

Table 2. Predicted and actual execution times.

	Predicted Exec. Time (s)	Empirical Exec. Time (s)	Refined Exec. Time (s)
Base System	1662.9	187.80 ± 20.02	1522.3
CBR-CNP	133.9	53.96 ± 7.75	117.04

Table 2 lists the mean execution times for both the integrated and base systems. The models predict that the execution time for approximately the same mission ($n = 2, g = 11, h = 11, \tau = 2$) will require about 12.4 times the execution time on the base system than on the integrated CBR-CNP system. Empirical results indicate that the actual execution times are less for both systems (based on five of six subjects). These experiments reveal that the execution time on the base system requires approximately 3.5 times the execution time compared to the integrated CBR-CNP system.

There are several possible reasons for the discrepancy between the predicted and empirical results. First, some primitive operator execution times may not be correct for this particular experiment. Gong and Kieras found that mouse movements are more accurately estimated from Fitts' Law than by the Keystroke-Level Model (KLM) time of 1.2 sec used here [11]. Second, the base system model does not assume that the user will use shortcuts although some subjects did. Third, neither model factors in the performance gains associated with repeatedly constructing the same missions. Regardless of the precise gains in performance realized by one system over the other, the most important point is that the GOMS models accurately indicate the utility of the CBR-CNP interface changes.

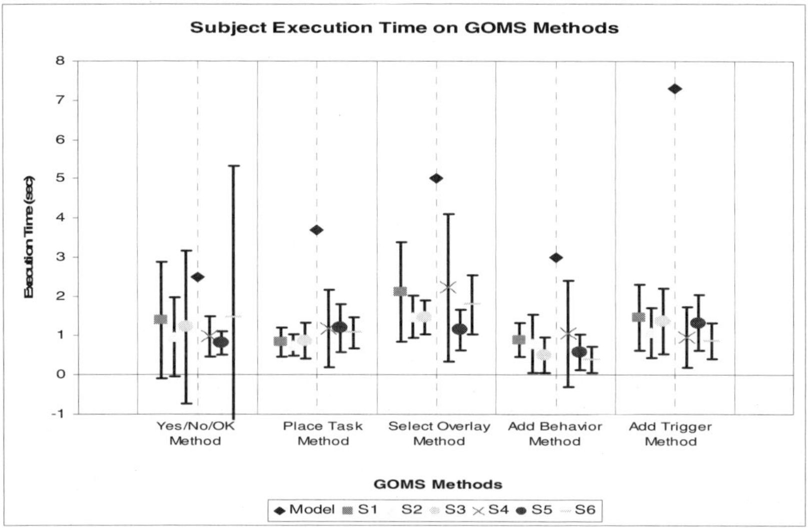

Fig. 2. A scatter diagram of the experimental execution time for expert users on five GOMS methods is depicted. Error bars represent 95% CI. The two rightmost methods are from the base model. The next two methods are from the CBR-CNP model and the leftmost method occurred in both models. The blue diamond on the dashed line depicts the execution time for each method predicted by the GOMS models. The left of each dashed line depicts the actual time for three experimenters. The right of each dashed line depicts the time for experimentally naïve expert users.

Figure 2 depicts the predicted execution times and actual execution times for several arbitrarily selected methods in the GOMS models. The independent variable represents the GOMS method selected and the dependent variable describes the subject's execution time. The `Yes/No/OK Method` (see appendix for GOMS methods) occurs in both GOMS models and experimental conditions. The `Place Task Method` and the `Select Overlay Method` occur in only the CBR-CNP model and experimental condition. Finally, the `Add Behavior Method` (presented in section 3) and the `Add Trigger Method` occur in the base system model and experimental condition. The dashed line denotes the execution time predicted by the models. The subjects to the left of each dashed line are also the experimenters. The subjects to the right of the dashed line are naïve subjects. The figure shows that the execution time predictions from our initial model are significantly greater then the actual expert execution times. As a result, we can now revisit the model and update the execution times for greater accuracy. Table 1 compares the initial model coefficients to refined model coefficients. Table 2 presents execution times based on these refined models. The predicted execution times for both refined models are closer to the actual execution times.

Several other points are also of interest. First, as indicated by figure 2, no experimenter bias is apparent. This is important because it increases the subject pool for potential user interface experiments. Hence, it appears that expert subjects can be drawn from the authors of the study itself, given the restrictions outlined above; possibly further reducing the challenge of user interface evaluation. Second, GOMS experiments are robust to experimental error. A data collection error occurred for one subject (*S4*) and another subject misunderstood the directions (*S3*). The data collection error (*S4*) resulted in elimination of this subject's mission execution time results (in Table 2) but did not affect the subject's data collected while completing the GOMS methods (data in Fig. 2). Subject 4's misunderstanding of the directions resulted in fewer data points for the `Add Behavior Method` (~150 versus ~220 normal). In spite of this experimental error, enough data was collected to produce statistically significant conclusions. Usability studies often face similar challenges and must completely exclude data from some subjects due to errors such as these. Nevertheless, because GOMS models are constructed from low-level user interface primitives, data from these subjects could still be salvaged.

Overall, both GOMS models and our experimental results indicate the value of the CBR-CNP user interface. Moreover, our modeling results and empirical results can be used for additional future interface design evaluations of this system.

6 Conclusions

This paper has investigated user interface modeling as a method for evaluating multi-robot interface design. We compared two GOMS models, one representing the base system with additional features for multi-robot multi-task mission generation and the other without. Our results indicate that these new multi-robot mission generation features substantially improve the usability of the *MissionLab* software. We intend to evaluate future interface designs using the same techniques, which may include the construction of a GOMS library of expert operator execution times. This should aid in the construction of more accurate future models.

We believe, and our case study has shown, that modeling techniques such as GOMS can play an important role in robotic interface development. Moreover, our work indicates that researchers can perform these techniques in a relatively inexpensive and timely manner. It is our sincere hope that other robotics researchers will consider the lessons described here, and in detail by HCI specialists [3], when designing user interfaces for critical robot applications operating in hazardous environments.

Acknowledgements

This research is funded as part of Navair's Intelligent Autonomy program. We would like to thank Ed Clarkson for lending his HCI expertise and Zsolt Kira, Lilia Moshkina, and Eric Martinson for assisting with this research.

References

1. J.A. Adams. (2002) Critical Considerations for Human Robot Interface Development, *Proceedings of 2002 AAAI Fall Symposium*, 1-8.

2. Y. Endo, D.C. MacKenzie, and R.C. Arkin. (2004) Usability Evaluation of High-Level User Assistance for Robot Mission Specification, IEEE Transactions on Systems, Man, and Cybernetics, **34**, 168-180.

3. D. Kieras (1994). *A guide to GOMS model usability evaluation using NGOMSL*, In M. Helander & T. Landauer (Eds.), The handbook of human-computer interaction. (Second Edition), North-Holland Amsterdam.

4. S.K. Card T.P. Moran, and A.P. Newell (1983) The Psychology of Human-Computer Interaction, Lawrence Erlbaum, Hillsdale, N.J.

5. D.C. MacKenzie and R.C. Arkin. (1998) Evaluating the Usability of Robot Programming Toolsets, International Journal of Robotics Research, **17**(4), 381-401.

6. B.E. John and D.E. Kieras. (1996) Using GOMS for User Interface Design and Evaluation: Which Technique?, ACM Transactions on Computer-Human Interaction, **3**(4), 287-319.

7. H A Yanco, J. L. Drury, and J. Scholtz. (2004) Beyond usability evaluation: Analysis of human-robot interaction at a major robotics competition, *Journal of Human-Computer Interaction*, 19, 117-149.

8. B.E. John and D.E. Kieras. (1996) The GOMS Family of User Interface Analysis Techniques: Comparison and Contrast, ACM Transactions on Computer-Human Interaction, **3**(4), 320-351.

9. D.C. MacKenzie, R.C. Arkin, and J.M. Cameron. (1997) Multiagent Mission Specification and Execution, Autonomous Robots, **4**, 29-52.

10. M.B. Dias, R.M. Zlot, N. Kalra and A. Stentz. (2005) Market-Based Multirobot Coordination: A Survey and Analysis, Carnegie Mellon University Tech Report CMU-RI-TR-05-13.

11. Gong, R., and Kieras, D. (1994) A Validation of the GOMS Model Methodology in the Development of a Specialized, Commercial Software Application. In *Proceedings of CHI,* Boston, MA, USA, pp. 351-357.

Appendix

```
Method for goal: select yes-no-ok                          KLM op Time(s)
 Step 1: Locate yes-no-ok button                               M       1.2
 Step 2: Move cursor to yes-no-ok button                       P       1.1
 Step 3: Click mouse button                                    BB      0.2
 Step 4: Return with goal accomplished                         Total   2.5
                                                       Refined Total 1.14
Method for goal: place task                                KLM op Time(s)
 Step 1: Think-of placement point                              M       1.2
 Step 2: Move cursor to placement point                        P       1.1
 Step 3: Click mouse button                                    BB      0.2
 Step 4: Verify that placement point is correct                M       1.2
 Step 5: Return with goal accomplished                         Total   3.7
                                                       Refined Total 0.91
Method for goal: select overlay                            KLM op Time(s)
 Step 1: Locate name in the file list box                      M       1.2
 Step 2: Move cursor to the file name location                 P       1.1
 Step 3: Click mouse button                                    BB      0.2
 Step 4: Locate ok button                                      M       1.2
 Step 5: Move cursor to ok button                              P       1.1
 Step 6: Click mouse button                                    BB      0.2
 Step 7: Forget overlay name and return                        Total   5.0
                                                       Refined Total 1.82
Method for goal: add trigger                               KLM op Time(s)
 Step 1: Locate trigger icon                                   M       1.2
 Step 2: Move cursor to trigger icon location                  P       1.1
 Step 3: Click mouse button                                    BB      0.2
 Step 4: Locate trigger tail behavior location                 M       1.2
 Step 5: Move cursor to trigger tail behavior                  P       1.1
 Step 6: Press mouse button down                               B       0.1
 Step 7: Locate trigger tip behavior                           M       1.2
 Step 8: Move cursor to tip behavior                           P       1.1
 Step 9: Release mouse button                                  B       0.1
 Step 10: Return with goal accomplished                        Total   7.3
                                                       Refined Total  0.87
```

Index